Essential Physics for Cambridge IGCSE® Workbook

For the updated syllabus

Sarah Lloyd

Oxford excellence for Cambridge IGCSE®

OXFORD UNIVERSITY PRESS

OXFORD
UNIVERSITY PRESS

Great Clarendon Street, Oxford, OX2 6DP, United Kingdom

Oxford University Press is a department of the University of Oxford. It furthers the University's objective of excellence in research, scholarship, and education by publishing worldwide. Oxford is a registered trade mark of Oxford University Press in the UK and in certain other countries

© Oxford University Press 2016

The moral rights of the authors have been asserted

First published in 2016

All rights reserved. No part of this publication may be reproduced, stored in a retrieval system, or transmitted, in any form or by any means, without the prior permission in writing of Oxford University Press, or as expressly permitted by law, by licence or under terms agreed with the appropriate reprographics rights organization. Enquiries concerning reproduction outside the scope of the above should be sent to the Rights Department, Oxford University Press, at the address above.

You must not circulate this work in any other form and you must impose this same condition on any acquirer

British Library Cataloguing in Publication Data
Data available

978-0-19-837469-5

3 5 7 9 10 8 6 4

Paper used in the production of this book is a natural, recyclable product made from wood grown in sustainable forests. The manufacturing process conforms to the environmental regulations of the country of origin.

Printed in India by Manipal Technologies Limited, Manipal

Acknowledgements

®IGCSE is the registered trademark of Cambridge International Examinations.

The publishers would like to thank the following for permissions to use their photographs:

Cover image: Joe McBride/Corbis.

Artwork by Q2A Media Services Pvt. Ltd. and OUP.

Although we have made every effort to trace and contact all copyright holders before publication this has not been possible in all cases. If notified, the publisher will rectify any errors or omissions at the earliest opportunity.

Links to third party websites are provided by Oxford in good faith and for information only. Oxford disclaims any responsibility for the materials contained in any third party website referenced in this work.

Introduction

This workbook is designed to accompany the *Essential Physics for IGCSE* student book. It is designed to help you develop the skills you need in order to help you do well in your IGCSE Physics examination. The book follows the order of the chapters in *Essential Physics for IGCSE*. Each page of questions provides additional questions related to each double page in the student book.

The questions focus on the areas you need to know about for your exam:

- Knowledge (memory work) and understanding (applying your knowledge to answer questions about familiar or unfamiliar situations).
- Handling information from data, tables, and graphs.
- Solving problems (including equations and calculations).
- Experimental skills and investigations.

The first 15 units include a range of question types that you might come across in your physics examinations:

- Choosing words to complete sentences: you are usually given a list of words to choose from. This will help you learn and remember key facts.
- Putting statements in the correct order or selecting the correct statement from a list.
- Some questions ask you to interpret data from diagrams, graphs, and tables. Others ask you to interpret the results of investigations that may be unfamiliar.
- Some pages include questions involving extended answers. These will help you organise your arguments and understand the depth of answer that is needed.

Other important features of this workbook that should help you succeed in physics include:

- An introductory Language Lab section in each of the first 15 units, which focuses on scientific words. These are often placed in a particular context. Examples include fill-in-the-gap exercises, word searches, and crosswords.
- A unit focusing on language and the importance of identifying key words in questions. This includes vocabulary practice as well as practice in reading and analysing questions. It also includes a glossary to help you understand important terms.
- A unit focusing on how to make the most of revision time through active revision and mind mapping.
- A unit on mathematics for physics. This includes practice in writing formulae, rearranging expressions, working through calculations, and drawing graphs.
- A unit on practical aspects of physics including planning an experiment (the selection of apparatus and materials and working safely), measuring, recording data, and drawing graphs. It also includes analysis of results and evaluation. This is followed by a unit suggesting how these aspects of practical physics can be applied to projects.
- A selection of IGCSE-style questions of the type that are set in the theory papers will help you to see connections between different parts of the syllabus.
- Full answers are given to all the questions.

We hope that the range of differing exercises in this workbook will help you develop your skills in and understanding of physics and help you succeed in this subject.

Contents

1 Motion
1.1	Making measurements	2
1.2	Distance–time graphs	3
1.3	More about speed	4
1.4	Acceleration	5
1.5	More about acceleration	6
1.6	Free fall	7
	Multiple choice questions	8

2 Forces and their effects
2.1	Mass and weight	9
2.2	Density	10
2.3	Force and shape	11
2.4	Force and motion	12
2.5	More about force and motion	13
2.6	Momentum	14
2.7	Explosions	15
2.8	Impact forces	16
	Multiple choice questions	17

3 Forces in equilibrium
3.1	Moments	18
3.2	Moments in balance	19
3.3	The principle of moments	20
3.4	Centre of mass	21
3.5	Stability	22
3.6	More about vectors	23
	Multiple choice questions	24

4 Energy
4.1	Energy transfers	25
4.2	Conservation of energy	26
4.3	Fuel for electricity	27
4.4	Nuclear energy	28
4.5	Energy from wind and water	29
4.6	Energy from the Sun and the Earth	30
4.7	Energy and work	31
4.8	Power	32
	Multiple choice questions	33

5 Pressure
5.1	Under pressure	34
5.2	Pressure in a liquid at rest	35
5.3	Pressure measurements	36
5.4	Solids, liquids, and gases	37
5.5	More about solids, liquids, and gases	38
5.6	Gas pressure and temperature	39
5.7	Evaporation	40
5.8	Gas pressure and volume	41
	Multiple choice questions	42

6 Thermal physics
6.1	Thermal expansion	43
6.2	Thermometers	44
6.3	More about thermometers	45
6.4	Thermal capacity	46
6.5	Change of state	47
6.6	Specific latent heat	48
6.7	Heat transfer (1): thermal conduction	49
6.8	Heat transfer (2): convection	50
6.9	Heat transfer (3): infrared radiation	51
6.10	Heat transfer at work	52
	Multiple choice questions	53

7 Waves
7.1	Wave motion	55
7.2	Transverse and longitudinal waves	56
7.3	Wave properties (1): reflection and refraction	57
7.4	Wave properties (2): diffraction	58
	Multiple choice questions	59

8 Light
8.1	Reflection of light	60
8.2	Refraction of light	61
8.3	Refractive index	62
8.4	Total internal reflection	63
8.5	The converging lens	64
8.6	Applications of the converging lens	65
8.7	Electromagnetic waves	66
8.8	Applications of electromagnetic waves	67
	Multiple choice questions	68

9 Sound
9.1	Sound waves	70
9.2	Properties of sound	71
9.3	The speed of sound	72
9.4	Musical sounds	73
	Multiple choice questions	74

10 Magnetism
10.1	Magnets	75
10.2	Magnetic fields	76
10.3	More about magnetic materials	77
	Multiple choice questions	78

11 Electric charge
11.1	Static electricity	79
11.2	Electric fields	80
11.3	Conductors and insulators	81
11.4	Charge and current	82
	Multiple choice questions	83

Contents

12 Electrical energy

12.1	Batteries and cells	84
12.2	Potential difference	85
12.3	Resistance	86
12.4	More about resistance	87
12.5	Electrical power	88
	Multiple choice questions	89

13 Electric circuits

13.1	Circuit components	90
13.2	Series circuits	91
13.3	Parallel circuits	92
13.4	More about series and parallel circuits	93
13.5	Sensor circuits	94
13.6	Switching circuits	95
13.7	Logic circuits	96
13.8	Logic circuits in control	97
13.9	Electrical safety	98
13.10	More about electrical safety	99
	Multiple choice questions	100

14 Electromagnetism

14.1	Magnetic field patterns	102
14.2	The motor effect	103
14.3	The electric motor	104
14.4	Electromagnetic induction	105
14.5	The alternating current generator	106
14.6	Transformers	107
14.7	High-voltage transmission of electricity	108
	Multiple choice questions	109

15 Radioactivity

15.1	Observing nuclear radiation	110
15.2	Alpha, beta, and gamma radiation	111
15.3	The discovery of the nucleus	112
15.4	More about the nucleus	113
15.5	Half-life	114
15.6	Radioactivity at work	115
	Multiple choice questions	116

16 Language focus

Glossary	118
Key word exercises	122

17 Revision

Revision tips	127
Revision checklists	129
Mind maps	131

18 Practical physics

Practical skills	135

19 Mathematics for physics

	144

20 Exam-style questions

	148

21 Project ideas

Answers	156
Data sheet	171

Motion

1.1 Making measurements

Language lab

Anagrams: unjumble the key words.

love mu ....

insincerely mud rag ....

muter reel ....

pock colts .... [4]

1. In international athletics competitions, such as the Olympics, it is essential to time races as accurately and precisely as possible.

   a. How are the races timed?

   ....

   ....

   .... [2]

   b. Why is this method of timing particularly important for short races such as the 100 metres?

   ....

   ....

   .... [2]

   c. Why would it be difficult to time a 100 metres race using a stop clock?

   ....

   .... [1]

2. a. You are asked to find, as accurately as possible, the volume of a pebble with approximate volume 30 cm³, using a water displacement method. Which size measuring cylinder would you choose to measure the displaced water?

   | 10 ml | 50 ml | 100 ml | 1000 ml | [1] |

   b. i. A student is given a block of wood with approximate dimensions 2 cm by 1 cm by 6 cm. Describe how he can find the volume of the block using a metre rule.

   ....

   ....

   ....

   ....

   .... [3]

   ii. Name a measuring instrument he could use to improve the precision of his measurements.

   .... [1]

Motion

1.2 Distance–time graphs

Language lab

Match the beginnings and endings of sentences.

Beginnings:

If an object moves at a steady speed it covers

The average speed for a journey can be found

The gradient of a distance–time graph is equal

Endings:

by dividing the total distance by the time taken.

to the speed of the object.

the same distance every second. [3]

1. a. Use the data to plot a distance–time graph for a person's journey.

time / s	0	100	200	300	400	500
distance / m	0	160	320	480	640	640

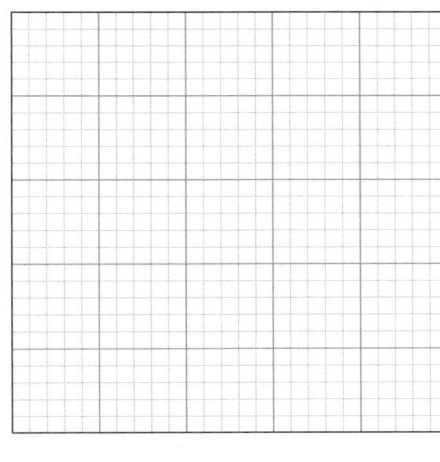

[4]

b. Use the graph to find the distance travelled in 150 s.

.... [1]

c. What physical quantity is given by the gradient of a distance–time graph?

.... [1]

d. i. Describe the motion of the person between 0 and 200 s.

....
.... [1]

ii. Describe the motion of the person between 400 s and 500 s.

....
.... [1]

2. Describe the journeys represented by the following distance–time graphs.

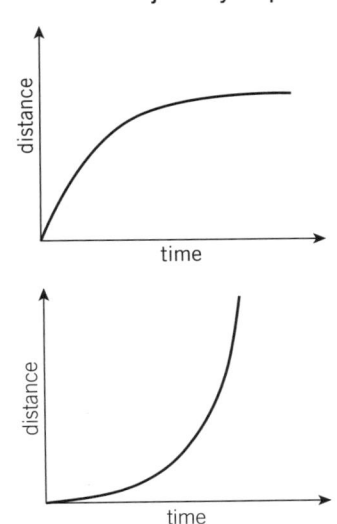

a. ....
....
....
.... [3]

b. ....
....
....
.... [3]

3

Motion

1.3 More about speed

Language lab

Describe your journey home from school or college. Include these key words: **speed**, **distance**, and **time**.

....

....

.... [3]

1. a. Using symbols, write down the equation for calculating speed.

   [1]

   b. i. Calculate the speed of a girl walking if she travels 100 m in 50 s.

      speed = ... m/s [1]

      ii. Calculate the average speed of a car that travels 2 km in 2 minutes.

      speed = ... km/h [2]

      iii. An athlete moving at 4.5 m/s travels a distance of 0.09 km. How long does this take?

      time = ... s [2]

   c. On a particular journey, a cyclist travels at a speed of 10 m/s for 35 minutes, rests for 5 minutes, and then travels at a speed of 8 m/s for 55 minutes.

      i. What was the total time for his journey?

      time = ... minutes [1]

      ii. What distance did he travel?

      distance = ... m [3]

4

Motion

1.4 Acceleration

Language lab

Anagrams: unjumble the key words.

conceal irate .... insect ad ....

city love .... emit ....

elected ear .... deeps .... [6]

1. a. Write down the equation for calculating acceleration. [1]

    b. Calculate the acceleration of a cyclist, who increases her velocity from 5 m/s to 7 m/s in 0.5 s.

    acceleration = ... [3]

    c. How long does it take for a train to increase its velocity from 10 m/s to 40 m/s if it accelerates at 3 m/s²?

    time = ... [3]

    d. A car, initially travelling at 6 m/s, accelerates at 4 m/s² for 2.5 s. What is its final velocity?

    velocity = ... [3]

Motion

1.5 More about acceleration

Language lab

Match the beginnings and endings of sentences.

Beginnings:

If the velocity of a car is increasing this means

When a cyclist applies his brakes the bicycle

The area under a velocity–time graph is equal

Endings:

will decelerate (slow down).

to the distance travelled.

it is accelerating.

1. a. i. What physical quantity is given by the area under a velocity–time graph?

   .... [1]

   ii. What is the difference between velocity and speed?

   ....

   .... [2]

   b. For the following velocity–time graph:

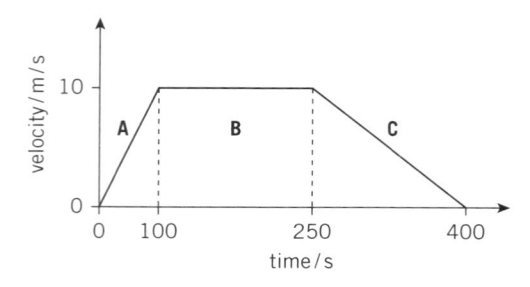

   i. Calculate the acceleration in sections **A**, **B**, and **C**.

   Section **A** .... [3]   Section **B** .... [2]   Section **C** .... [3]

   ii. Find the distance travelled in each section.

   Section **A** .... [3]   Section **B** .... [2]   Section **C** .... [2]

   iii. Calculate the total distance travelled.

   distance = ... [1]

Motion

1.6 Free fall

Language lab

Complete the crossword.

Across

3. Used to measure length
4. Used to measure the volume of a liquid
5. The unit of length
7. The unit of force
8. Calculated by multiplying length by width
10. Calculated by multiplying length by width by height
11. Used to measure time
14. Used to measure force

Down

1. The smallest measurement on a metre rule
2. The unit of acceleration
4. The unit of speed
6. The unit of time
12. Used to measure mass
13. The unit of energy

[14]

1. Two IGCSE students are carrying out an experiment to measure the acceleration due to gravity by dropping a double interrupt card through a light gate. The light gate automatically starts a timer as the card interrupts the light beam and stops the timer when the beam is no longer interrupted. The students input the width of the two sides of the card into the data logger.

   a. Explain how, using the two widths and the times for which they interrupt the beam, the data logger is able to calculate the acceleration due to gravity.

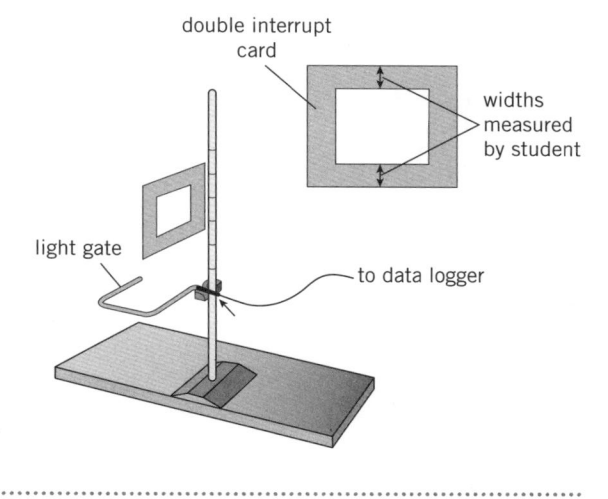

   ....
   ....
   ....
   ....
   ....
   .... [4]

   b. Student A says that as the height of drop of the double interrupt card is increased, the acceleration will increase. Student B disagrees and says that the height of drop will not matter and the acceleration will remain constant. Who is correct? Explain your answer.

   ....
   ....
   ....
   .... [3]

7

Motion — Multiple choice questions

1. What is the SI unit of length?

   A mm B cm
   C m   D km

2. What is the SI unit of time?

   A milliseconds B seconds
   C minutes   D hours

3. Which of the following is not an instrument used to measure length?

   A Vernier callipers B micrometer screw gauge
   C balance   D metre rule

4. What is the area of a piece of paper of dimensions 11.2 cm by 15.4 cm?

   A 172.48 m²   B 172.48 cm²
   C 0.17248 m² D 172.48 cm³

5. What is the volume in cm³ of a block of height 0.23 m, length 0.06 m, and width 0.15 m?

   A 2007 B 0.00207
   C 2070 D 0.0207

6. A toy car of volume 54 cm³ is placed into a 500 ml measuring cylinder, containing 250 ml of water. What is the new reading on the measuring cylinder?

   A 554 ml B 446 ml
   C 304 ml D 354 ml

7. A water wave takes 5.02 s to travel the length of a tank 3 times. What is the speed of the water wave if the tank is 1.5 m long?

   A 8.96 m/s   B 0.299 m/s
   C 0.299 cm/s D 0.896 m/s

8. Which of the following is the correct formula for calculating acceleration?

   A velocity = acceleration × time
   B acceleration = change in velocity ÷ time
   C acceleration = change in velocity × time
   D acceleration = time ÷ change in velocity

9. What is the acceleration of a runner who increases her velocity from 2 m/s to 8 m/s in 3 s?

   A 3.33 m/s² B 18 m/s²
   C 2.67 m/s² D 2 m/s²

Forces and their effects 2.1 Mass and weight

Language lab

Match the key words with their meanings.

mass	force on an object in a gravitational field
weight	the force on a mass of 1 kg due to gravity
gravitational field strength	SI unit of mass
kilogram	SI unit of weight
newton	the amount of matter in an object [5]

1. a. State the relationship between mass and weight, using words.

   ....

   .... [1]

   b. i. What is the SI unit of mass? .... [1]

   ii. What is the SI unit of weight? .... [1]

   c. Use the data to calculate the weight of a 60 kg person on each planet. [5]

Planet	Mercury	Venus	Earth	Mars	Jupiter
gravitational field strength / m/s²	3.76	9.04	9.81	3.77	23.6
weight of 60 kg person / N					

9

Forces and their effects — 2.2 Density

Language lab

Complete the sentences using the key words below.

| volume | balance | irregular | density | mass | measuring cylinder | rule |

The .... of an object can be calculated if its .... and .... are known. Mass can be found using a .... Volume can be found using a .... if the object is ...., or a .... if it is regular. [8]

1. a. Give two possible units for density. .... [2]

   b. Name a measuring instrument used to find the mass of an object. .... [1]

   c. i. Calculate the density of vegetable oil if 200 cm³ of oil has a mass of 180 g.

   density = ... [2]

   ii. Will the oil float on water (density 1 g/cm³)? Explain your answer.

   ....

   ....

   .... [2]

   d. Calculate the mass of 25 cm³ of mercury of density 13.5 g/cm³.

   mass = ... [2]

2. a. Using a ruler, measure the height and radius of the cylinder shown in the diagram. Using your measurements, calculate the volume of the cylinder.

   Height = ... [1]

   Radius = ... [1]

   Volume = ... [3]

   b. If the cylinder is made of iron of density 7.9 g/cm³, what is the mass of the cylinder?

   mass = ... [2]

Forces and their effects — 2.3 Force and shape

Language lab

Match the beginnings and endings of the sentences.

Beginnings:

Increasing the stretching force on a spring

The extension of a spring is calculated by

On a graph, the independent variable goes

Endings:

subtracting the original length from the new length.

on the x-axis.

increases the extension. [3]

1. a. Describe an experiment to investigate the effect of increasing the stretching force on the extension of a spring. Include a labelled diagram.

   ....

   ....

   ....

   ....

   ....

   .... [6]

   b. A student suggests a hypothesis that the extension of the spring is directly proportional to the stretching force. By setting up an experiment to investigate this hypothesis, she obtains the following results:

   Original length of the spring = 5.0 cm

force / N	0	0.5	1.0	1.5	2.0
length / cm	5.0	5.8	6.6	7.5	8.5
extension / cm					

   Complete the table and then plot a graph of extension vs force.

   [9]

   c. Do you agree with the student's hypothesis? Explain why, or why not.

   ....

   .... [3]

Forces and their effects

2.4 Force and motion

Language lab

Match the types of forces with their definitions.

friction	the force between two charged particles
weight	the force between two poles
air resistance	opposes motion when two surfaces rub together
magnetic	a type of frictional force, which opposes motion in a fluid
electrostatic	the force due to gravity [5]

1. a. Forces cause changes. Name three effects a force can have on an object.

   1 ....

   2 ....

   3 .... [3]

   b. Name two forces that oppose the motion of an object.

   1 ....

   2 .... [2]

2. Student A, student B, and student C are having a discussion about the effects of resultant forces on the motion of objects.

   Student A says:
   'If there is no resultant force on an object then the object will not move.'
   Student B says:
   'If there is no resultant force on an object, it will keep moving at a steady speed in a straight line.'
   Student C says:
   'All objects slow down eventually if there is no resultant force to keep them moving.'

   To what extent do you agree with the three statements?

   ....

   ....

   ....

   .... [3]

3. Calculate the resultant force in each case.

   a. → 3 N
      → 2 N

   b. 3.5 N ←   → 6.5 N

   c. 60 N ←   → 50 N

   .... [3]

12

Forces and their effects
2.5 More about force and motion

Language lab

Anagrams: unjumble the key words.

fluorescent art ....

coin rift ....

moon it ....

mentor we net .... [4]

1. A student wishes to test which materials produce the greatest frictional force. He wraps each material in turn around a wooden block and pulls the block, weighted with a 200 g mass, at a steady speed across a smooth surface using a newton meter as shown in the diagram below.

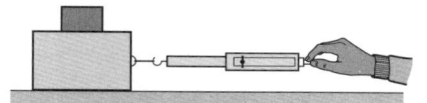

   The student obtains the following results:

material	force / N	force / N	average force / N
paper	0.8	0.7	
cotton cloth	0.9	1.0	
polythene bag	0.6	0.6	
sand paper	1.4	1.3	

   a. Calculate the average forces and enter them in the table. [1]

   b. What is the resolution of the newton meter? .... [1]

   c. Why did the student place the 200 g mass on the block?

   ....

   .... [1]

   d. Why was it important for him to pull the block with a steady speed?

   ....

   .... [2]

   e. Which material had the smallest frictional force? Draw a labelled diagram to explain why the frictional force is least for this material.

   [2]

   f. Another student repeats the experiment; holding the newton meter at an angle of 45° to the horizontal surface. How would you expect his results to compare to the results in the table above? Explain your answer.

   ....

   .... [2]

13

Forces and their effects — 2.6 Momentum

Language lab

Match the key words and their definitions.

momentum	change in momentum
impulse	speed in a certain direction
velocity	mass × velocity

[3]

1. a. What is the unit of momentum? [1]

   ....

   b. i. What is the momentum of a 1000 kg car moving at 20 m/s?

   momentum = ... [1]

   ii. The momentum of a cyclist is 100 kg m/s. If the mass of the cyclist and her bicycle is 70 kg, what is her velocity?

   velocity = ... [2]

   iii. What is the mass of a boy if he has a momentum of 200 kg m/s and is running at 4 m/s?

   mass = ... [2]

   c. i. A toy car of mass 0.5 kg increases its velocity from 1 m/s to 2.5 m/s. What impulse did it receive?

   impulse = ... [2]

   ii. A trolley of mass 1 kg travelling at 0.6 m/s receives an impulse of 1.6 kg m/s. What is the new velocity of the trolley?

   velocity = ... [3]

   iii. A force of 25 N acts on a 30 kg shopping trolley for 1.5 s. What is the velocity of the trolley if it was initially at rest?

   velocity = ... [3]

Forces and their effects — 2.7 Explosions

Language lab

Complete the sentences using the key words.

| recoils | bullet | momentum | speed | opposite | Conservation |

When a .... is fired from a gun, it travels at great ...., which means that it has a large .... The Principle of .... of Momentum predicts that the gun will have equal but .... momentum to the bullet. Therefore the gun .... [6]

1. A bullet of mass 30 g leaves a gun of mass 700 g at a velocity of 200 m/s.

   a. What is the momentum of the bullet and gun before the explosion?

   momentum = ... [1]

   b. What is the momentum of the bullet after the explosion?

   momentum = ... [3]

   c. What is the momentum of the gun after the explosion?

   momentum = ... [1]

   d. What is the recoil velocity of the gun?

   velocity = ... [2]

Forces and their effects — 2.8 Impact forces

Language lab

Anagrams: unjumble the key words.

rec of .... net mom mu ....

lime sup .... emit .... [4]

1. a. i. Write down the formula for calculating impulse.

   .... [1]

   ii. What is the unit of impulse? .... [1]

   b. i. In a collision a car of mass 1000 kg slows from 25 m/s to 15 m/s. Calculate the impulse on the car.

   impulse = ... [2]

   ii. A cyclist and bicycle receive a backwards impulse of 250 kg m/s. The cyclist and her bicycle originally had a momentum of 325 kg m/s. What is their new momentum?

   momentum = ... [2]

   c. During a collision a force of 82 000 N acts on a car for 0.03 s. Calculate the change in momentum of the car.

   change in momentum = ... [2]

   d. i. A car of mass 1200 kg, which is travelling at 11 m/s, collides with another car and stops in 0.1 s. Calculate the force on the car.

   force = ... [3]

   ii. Seat belts in cars are made of slightly stretchy fabric. Explain how this helps to reduce the force on a passenger during a collision.

   ....

   ....

   .... [3]

2. a. Calculate the force exerted on an object if it accelerates at 5 m/s^2 and has a mass of 0.5 kg.

   force = ... [2]

   b. Calculate the acceleration produced if a force of 3000 N acts on a car of mass 900 kg.

Forces and their effects — Multiple choice questions

1. What is the resultant force acting on a ball of mass 300 g if its acceleration is 4.5 m/s²?

    A 1350 N   B 1.35 N

    C 66.7 N   D 15 N

2. What is the momentum of a car of mass 1000 kg travelling with a velocity of 24 km/h?

    A 400 000 kg m/s   B 6700 kg m/s

    C 24 000 kg m/s   D 2800 kg m/s

3. A bullet of mass 10 g leaves a gun of mass 500 g at a velocity of 200 m/s. What is the recoil velocity of the gun?

    A 4 m/s   B 40 m/s

    C 10 m/s   D 0.1 m/s

4. Which of the following is not a vector quantity?

    A kinetic energy   B force

    C acceleration   D velocity

5. A metal block of dimensions 1.5 cm by 2.3 cm by 4.5 cm has a mass of 42 g. What is the density of the metal?

    A 2.7 g/cm³   B 652 g/cm³

    C 3.6 g/cm³   D 3.2 g/cm³

6. A spring has a spring constant of 32 N/m. It extends by 2.0 cm when a force is applied. What is the value of the force?

    A 0.32 kg   B 0.32 N

    C 0.64 N   D 0.064 kg

Forces in equilibrium — 3.1 Moments

Language lab

Complete the sentences using the keywords.

    pivot   newtons   mass   turning   perpendicular   metres

Moment is equal to .... times .... distance from the .... ,

where force is measured in .... and distance is measured in ....

A moment causes a .... effect on an object. [6]

1. a. A force of 4.5 N acts at a distance of 1.5 m from a pivot. Find the moment of the force about the pivot.

   moment = ... [2]

   b. A force acting at a distance of 0.8 m from a pivot produces a turning moment of 1.6 N m. Calculate the size of the force.

   force = ... [2]

2. a. In the following examples, the objects are in equilibrium. Find the unknown values.

   i. (10 N upward at distance d left of pivot; 3 m to the right; 6 N down at left end; 4 N down at right end)

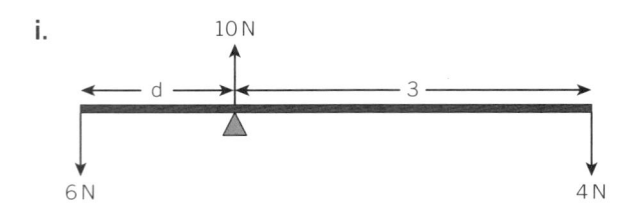

   d = ...

   ii. (F_1 upward at pivot; 1.5 m to left, 1 m to right; 5 N down at left end; F_2 down at right end)

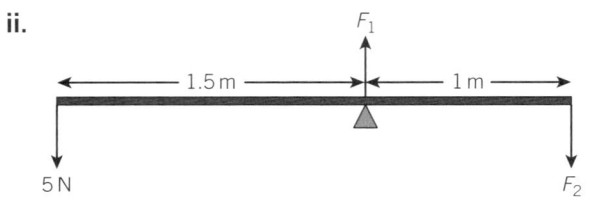

   F_1 = ...

   F_2 = ... [3]

Forces in equilibrium — 3.2 Moments in balance

Language lab

Match the key words with their definitions.

spring balance	the point about which an object can turn
equilibrium	a meter used to measure weight
net moment	when an object has no net moment or force acting on it
pivot	sum of all the clockwise and anticlockwise moments

[4]

b. You are given a metre rule, a spring balance, and a 9 N weight in the arrangement shown in the diagram (right).

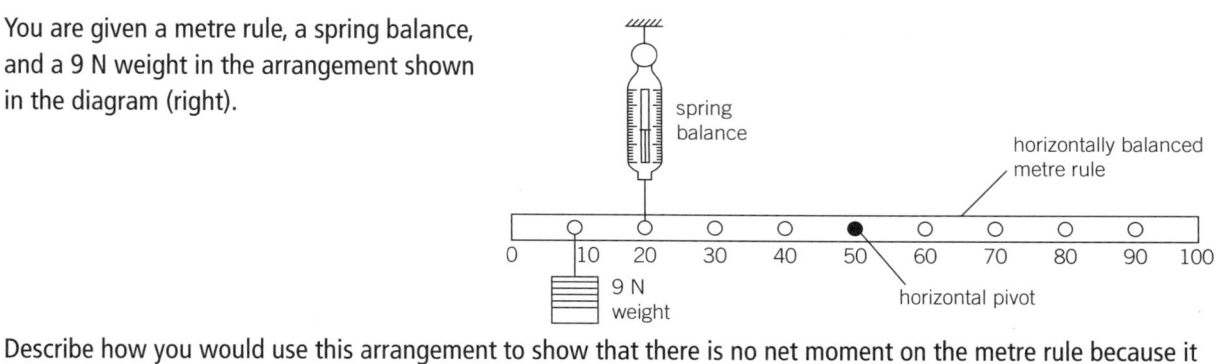

Describe how you would use this arrangement to show that there is no net moment on the metre rule because it is in equilibrium.

....

....

....

....

.... [4]

2. The diagram below shows a crane, which is used to lift and move heavy loads on construction sites.

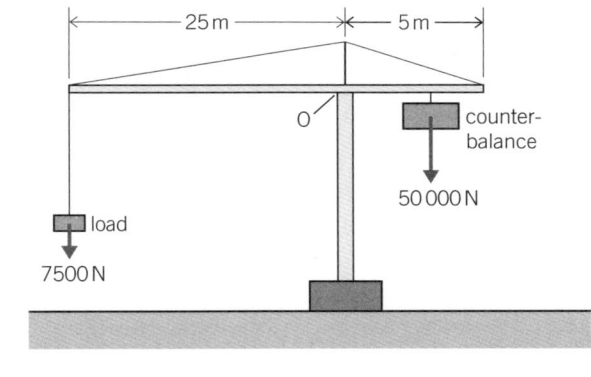

   a. Calculate the turning moment due to the load. Give the correct unit.

   Moment = ... [3]

b. If the crane is in equilibrium, state the turning moment due to the counterbalance. .... [1]

c. How far from O should the counterbalance be placed in order to balance the crane?

   Distance = ... [2]

d. Calculate the maximum load that the crane can safely lift with this counterbalance.

   Load = ... [2]

19

Forces in equilibrium — 3.3 The principle of moments

Language lab

Complete the word search.

Words may go downwards, upwards, forwards, backwards, or diagonally.

- anticlockwise
- balance
- clockwise
- distance
- equilibrium
- force
- horizontal
- metre
- moment
- newton
- perpendicular
- pivot

E	C	N	A	T	S	I	D	F	C	H	N	N	M	N
K	Q	G	E	M	C	L	Z	L	O	O	Z	E	E	V
Q	D	U	K	C	D	I	O	X	T	R	S	T	T	Z
W	W	C	I	F	N	C	D	W	N	I	C	N	R	C
R	S	E	G	L	K	A	E	B	W	Z	D	E	E	N
O	J	W	M	W	I	N	L	K	G	O	N	M	U	W
D	B	U	I	E	T	B	C	A	F	N	V	O	T	Y
B	E	S	G	V	E	O	R	T	B	T	I	M	U	G
V	E	G	T	I	L	C	K	I	V	A	U	T	K	E
D	A	W	Y	C	N	X	H	U	U	L	Z	Z	D	G
M	S	P	I	V	O	T	I	V	R	M	W	U	S	L
J	I	T	N	B	X	X	M	J	M	V	Q	I	U	Q
N	N	V	R	U	X	R	L	U	B	U	V	W	R	U
A	L	F	A	U	S	E	E	H	M	H	R	O	O	Q
P	E	R	P	E	N	D	I	C	U	L	A	R	H	D

[12]

1. The diagram below shows a hammer being used to remove a nail.

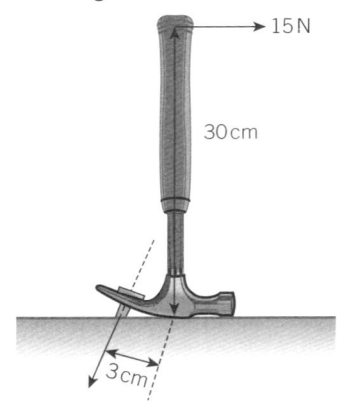

   → 15 N

   30 cm

   3 cm

   a. Calculate the turning moment due to the force of the person's hand. Give the unit.

   Moment = ... [3]

   b. Calculate the force on the nail.

   Force = ... [2]

20

Forces in equilibrium

3.4 Centre of mass

Language lab

Anagrams: unjumble the key words.

net mom ....

rublieumiqi ....

i op tv ....

cafe monsters .... [4]

1. For the following shapes, mark on the centre of mass.

[4]

2. A student wants to find the centre of mass of a piece of thin card in the shape of a tree.
Describe, with the aid of a diagram, how she can do this using a small weight tied to a piece of string and a pin.

....
....
....
....
....
....
....
.... [5]

Forces in equilibrium — 3.5 Stability

Language lab

Complete the sentences using the key words.

weight   topple   stable   moment   mass   base

An object is .... if the centre of .... lies vertically above its base so that the line of action of the .... acts through the .... If the object is tilted so that the line of action of the weight no longer acts through the base, there will be a net .... on the object and it will .... [6]

1. a.

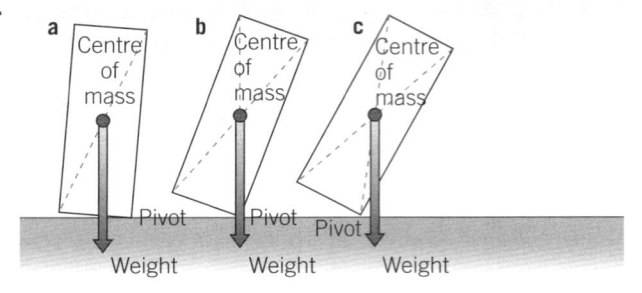

   i. In the diagram above, which one of the blocks will topple? .... [1]

   ii. Explain why.

   ....

   ....

   .... [2]

   b. i. Explain why a vase that is half filled with water is less likely to topple over than one that is empty.

   ....

   ....

   .... [3]

   ii. Buses in which most of the passengers are on the top deck are less stable than ones in which the passengers are seated on the lower deck. Explain why.

   ....

   ....

   .... [3]

   c. A tractor has a wide wheel base. Explain how this helps to increase the tractor's stability.

   ....

   ....

   .... [3]

Forces in equilibrium

3.6 More about vectors

Language lab

Complete the sentences using the key words.

| resultant | diagram | size | force | vector | direction |

The .... force is the .... sum of two or more forces. This takes into account both the .... and .... of the forces. The resultant .... can be determined by the use of an accurately drawn and scaled vector .... [6]

1. For the following examples, use a graphical method to find the resultant force.

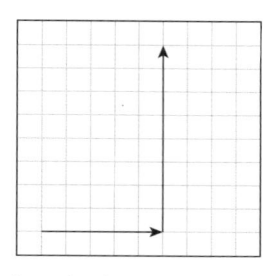

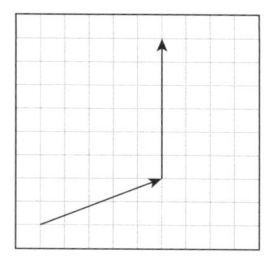

   length of one square represents 2 N [4]

2. A boat has run aground and two tug boats are pulling it ashore.

   Tug A pulls to the west with a force of 4000 N. Tug B pulls south with a force of 2800 N.

   In the space below draw a scale diagram where 1 cm represents 1000 N and use it to find the resultant force of tug A and tug B on the boat.

Forces in equilibrium — Multiple choice questions

1. Calculate the force required to cause a turning moment of 30 N m about a pivot at a distance of 1.2 m from the pivot.

    A 36 N

    B 0.36 N

    C 0.25 N

    D 25 N

2. What is the turning moment of a force of 12 N acting at a perpendicular distance of 0.3 m from a pivot?

    A 40 Nm

    B 4 Ncm

    C 3.6 Nm

    D 0.025 Nm

3. At what perpendicular distance from the pivot does a force of 1700 N act if the turning moment is 34 000 Nm?

    A 50 m

    B 20 m

    C 0.05 m

    D 200 m

4. Which of the following is a vector quantity?

    A mass

    B weight

    C density

    D energy

Energy

4.1 Energy transfers

Language lab

Complete the word search.

Words may go downwards, upwards, forwards, backwards, or diagonally.

- chemical
- electrical
- energy
- internal
- kinetic
- light
- nuclear
- sound
- strain
- thermal

[10]

E	N	R	F	A	X	C	G	E	I	C	R	L	B	D
U	L	P	R	C	B	R	D	N	X	H	A	I	H	L
Y	N	E	Q	C	Z	N	T	J	V	E	E	G	H	D
Z	U	B	C	I	T	E	N	I	K	M	L	H	F	A
I	T	Q	S	T	R	O	M	D	B	I	C	T	D	B
Z	J	L	E	N	R	M	E	M	R	C	U	M	Y	A
W	W	G	A	C	B	I	Y	E	S	A	N	V	T	S
U	Y	L	R	M	Q	G	C	E	K	L	M	P	J	V
M	G	R	K	P	R	I	E	A	W	T	V	H	Z	E
J	S	I	X	E	K	E	B	M	L	H	D	H	S	B
P	A	O	N	M	A	Y	H	G	E	F	E	R	Y	X
Z	Q	E	U	D	N	E	S	T	R	A	I	N	J	H
E	B	K	D	N	X	I	J	F	A	Z	T	H	O	Q
H	J	F	A	J	D	T	J	T	S	Q	N	I	G	I
Q	M	M	T	F	O	W	W	H	K	M	T	Q	I	Z

1. Match the types of energy with their descriptions.

chemical	stored in batteries and food
gravitational potential	energy an object has due to its movement
kinetic	stored energy that can be released by fusion
thermal	total kinetic and potential energies of all the particles in an object
strain	gained as an object is moved away from the Earth
sound	emitted by very hot objects
light	released when the temperature of a hot object decreases due to a decrease in its internal energy
nuclear	produced by vibrating objects
electrical	stored when an object changes shape
internal	carried by moving charges in a circuit

[10]

Energy
4.2 Conservation of energy

Language lab

Anagrams: unjumble the key words.

meal chic	....	lactic reel	....
cite ink	....	art helm	....
nod us	....	ran clue	....
a palatial ore voting tint	....		[7]

1. a. Complete the energy transformations.

light bulb	electrical	⟶	
electric iron		⟶	thermal
microphone		⟶	
waterfall		⟶	
catapult		⟶	
electric mixer		⟶	

    [10]

    b. State the principle of conservation of energy.

    ....

    .... [2]

2. a. Complete the energy diagrams.

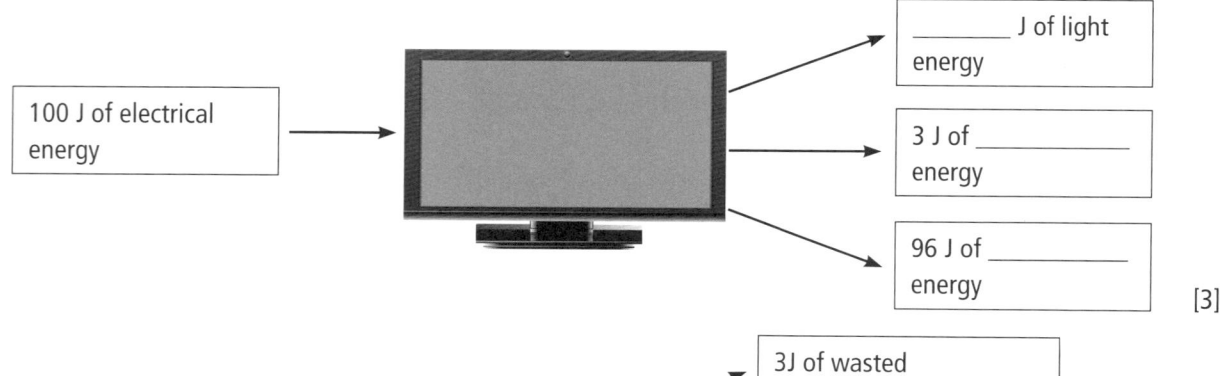

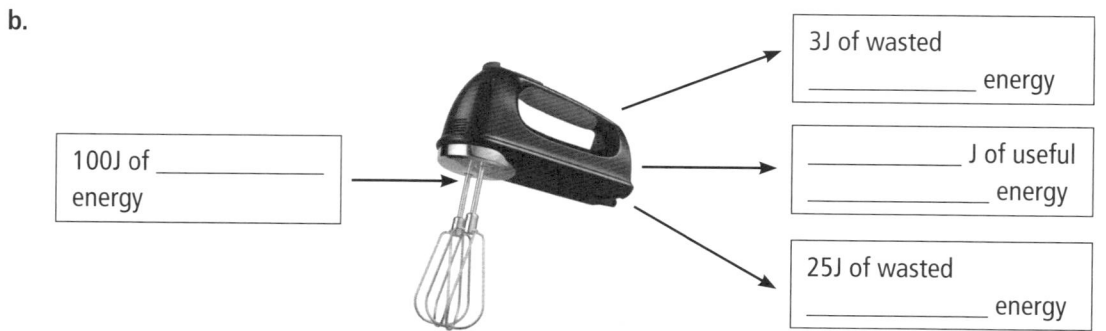

    100 J of electrical energy ⟶ [TV] ⟶ _____ J of light energy
    ⟶ 3 J of _____ energy
    ⟶ 96 J of _____ energy

    [3]

    b.

    100J of _____ energy ⟶ [mixer] ⟶ 3J of wasted _____ energy
    ⟶ _____ J of useful _____ energy
    ⟶ 25J of wasted _____ energy

Energy

4.3 Fuel for electricity

Language lab

Match the beginnings of the sentences to the endings.

Beginnings:

In a coal-fired power station, chemical energy

Efficiency is the ratio of output energy

Burning coal produces carbon dioxide which

Burning coal produces sulfur dioxide which

Some of the heat energy produced when coal is

The efficiency of a coal-fired power station is

Endings:

causes acid rain.

contributes to the greenhouse effect.

stored in coal is converted to electrical energy.

to input energy.

typically about 30%.

burnt is dissipated into the surroundings. [6]

1. a. Complete the boxes below to show the energy transformations that occur in a coal-fired power station. [6]

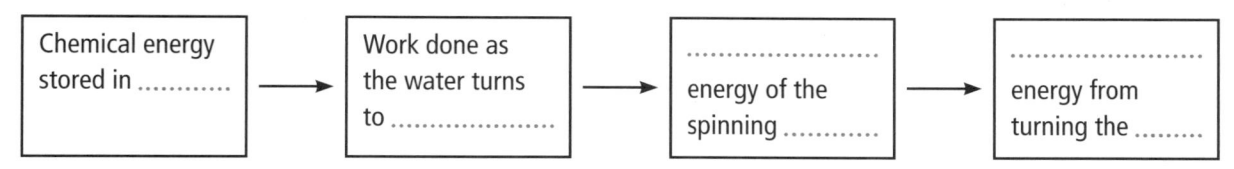

   Chemical energy stored in .... → Work done as the water turns to .... → .... energy of the spinning .... → .... energy from turning the ....

   b. How would the energy transformations above change for a nuclear power station?

   .... [1]

   c. i. The efficiency of a fossil fuel power station is 30%. How much chemical energy must be supplied to the power station each second if its output power is 2000 MW?

   Chemical energy supplied = ... [3]

   ii. In what form is most of the energy wasted from the power station? .... [1]

   iii. Where does this wasted energy go? .... [1]

   d. Which two pollutant gases are produced by coal-fired power stations?

   .... [2]

Energy

4.4 Nuclear energy

Language lab

Complete the sentences with the missing words.

turbines   water   nuclear   kinetic   energy   generator   steam   fission

In a .... power station, thermal energy is produced through nuclear ....

The thermal energy is used to heat .... to produce ...., which turns the

.... The turbines then turn the ...., which converts .... to

electrical energy. [7]

1. 1 kg of uranium, which is used in nuclear power stations, can produce 2 million times as much electrical energy as 1 kg of coal burnt in a coal-fired power station.

    Explain how this reduces the amount of energy wasted in transporting fuel.

    ....

    ....

    .... [3]

Energy

4.5 Energy from wind and water

Language lab

Anagrams: unjumble the key words.

correctly hide .... clue ran ....

gather mole .... burn lie ale ....

fuses so fill .... wane rebel .... [6]

1. Complete the energy diagrams.

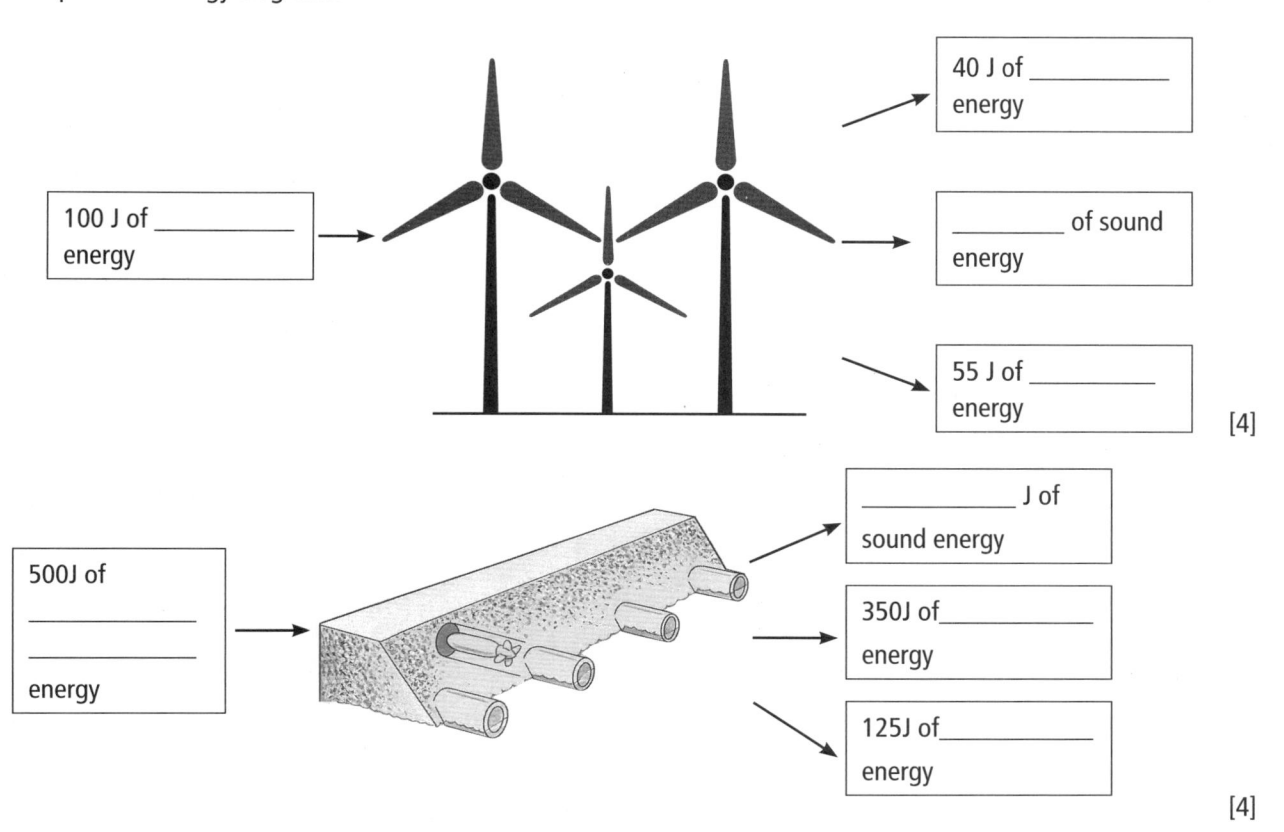

   100 J of _____ energy →

   → 40 J of _____ energy

   → _____ of sound energy

   → 55 J of _____ energy [4]

   500 J of _____ _____ energy →

   → _____ J of sound energy

   → 350 J of _____ energy

   → 125 J of _____ energy [4]

2. For each of the energy resources above, calculate the efficiency.

   Efficiency of wind turbines = ... [2]

   Efficiency of tidal barrage = ... [2]

3. Neither wind nor tidal power releases pollutant gases into the atmosphere but some people argue that they cause pollution in other ways. Explain how.

   ....

   ....

   .... [3]

29

Energy

4.6 Energy from the Sun and the Earth

Language lab

Match the key words to their definitions.

renewable	an energy resource that produces electrical energy when required
non-renewable	a means of producing electrical energy through energy transformations
reliable	a store of chemical energy formed millions of years ago
unreliable	a solar panel that transforms light energy to electrical energy
fuel	an energy resource that will eventually run out
energy resource	an energy resource that is not always available
photovoltaic cell	a source of chemical energy that is burnt to release energy as heat
fossil fuel	an energy resource that will not run out

[8]

1. a. Sort the energy resources into renewable and non-renewable.

   wind   wave   solar   coal   oil   geothermal   hydroelectric   nuclear   gas   biofuel   tidal

renewable	non-renewable

   [11]

   b. Sort the energy resources into those that are reliable (produce energy at a constant rate) and those that are unreliable (are affected by factors such as the weather).

   wind   wave   solar   coal   oil   geothermal   hydroelectric   nuclear   gas   biofuel   tidal

reliable	unreliable

   [11]

   c. Sort the energy resources into those that require fuel and those that do not.

   wind   wave   solar   coal   oil   geothermal   hydroelectric   nuclear   gas   biofuel   tidal

require fuel	do not require fuel

   [11]

Energy

4.7 Energy and work

Language lab

Use the key words to describe and explain the energy transformations that occur when a child slides down a slide.

kinetic   gravitational   potential   energy   friction   thermal

....

....

.... [6]

1. a. Calculate the kinetic energy of a 50 kg girl running at 4 m/s.

   kinetic energy = ... [3]

   b. A ball of mass 400 g, originally at rest, gains kinetic energy of 20 J. What is its velocity?

   velocity = ... [3]

   c. Calculate the change in gravitational potential energy as a 40 kg child jumps from a 0.5 m high wall.

   change in GPE = ... [2]

   d. Calculate the change in height of a parachutist of mass 70 kg if his change in gravitational potential energy is 105 000 J.

   change in height = ... [2]

2. a. Calculate the work done by a car engine that produces a force of 1500 N to accelerate the car through a distance of 200 m.

   work done = ... [2]

   b. A car accelerates through a distance of 50 m, with the engine doing 40 kJ of work. What is the force supplied by the engine?

   force = ... [3]

31

Energy

4.8 Power

Language lab

Anagrams: unjumble the key words.

we pro .... cape regent ....

cynic fee if .... factor in .... [4]

1. a. i. What is the unit of power? .... [1]

   ii. What is the formula relating energy and power?

   [1]

   b. i. What is the formula relating energy and efficiency?

   [1]

   ii. Why is efficiency expressed as a fraction or percentage without a unit?

   .... [1]

2. a. A television requires 9000 J of electrical energy to run it for 30 s. What is the power of the television?

   Power = ... [2]

   b. How much electrical energy is required to run a 3 kW oven for 15 minutes?

   Energy = ... [3]

   c. How long would a 25 W loudspeaker work for if it was supplied with 1 MJ of energy?

   Time = ... [2]

3. a. What is the efficiency of a television that produces 12 J of light and 48 J of sound for every 200 J of electrical energy?

   Efficiency = ... [2]

   b. What is the energy input to a device that is 80% efficient and produces and output of 50 W?

   Energy input = ... [2]

   c. A 60 W light bulb emits 400 J of light energy in 100 s. What is its efficiency?

   Efficiency = ... [3]

Energy — Multiple choice questions

1. What is the percentage efficiency of a television if for every 50 kJ of electrical energy input, 45 kJ is dissipated into the surroundings as heat?

    A 0.1

    B 10

    C 90

    D 0.9

2. Friction transforms kinetic energy into what type of energy?

    A electrical

    B thermal

    C gravitational potential

    D chemical

3. What is the useful energy transformation in a motor?

    A electrical to heat

    B electrical to sound

    C electrical to kinetic

    D kinetic to electrical

4. What is the kinetic energy of a car of mass 1000 kg travelling at a velocity of 20 m/s?

    A 2 kJ

    B 400 000 J

    C 40 kJ

    D 200 000 J

5. What is the gravitational potential energy gained by a car of mass 1200 kg when it climbs a hill of height 4 m?

    A 4.8 kJ

    B 48 000 kJ

    C 48 kJ

    D 4800 kJ

Pressure

5.1 Under pressure

Language lab

Use the key words to complete the sentences. Each word may be used more than once.

| newtons | area | pressure | pascals |

When a force acts over an ...., a .... is exerted on that area. Force is measured in .... and .... is measured in m², so that .... is measured in N/m² or .... [6]

1. a. i. What are the units of pressure if force is measured in newtons and area is measured in cm²?

   .... [1]

   ii. What are the units of pressure if force is measured in newtons and area is measured in m²?

   .... [1]

   b. i. Calculate the pressure under a block of mass 0.8 kg if it rests on a side of area 22 cm².

   pressure = ... [3]

   ii. Calculate the weight of an object if it exerts a pressure of 20 Pa over an area of 0.4 m².

   weight = ... [2]

   c. A brick of mass 1.5 kg and dimensions 10 cm by 10 cm by 20 cm rests on the ground. Calculate the maximum pressure in N/cm² that it can exert on the ground.

   pressure = ... [3]

Pressure

5.2 Pressure in a liquid at rest

Language lab

Match the beginnings of the sentences to their endings.

Beginnings:

The pressure in a liquid is

As the depth of a liquid increases

Pressure in a liquid is dependent on the

The pressure deep in the ocean trenches is great

Endings:

density of the liquid.

the same in all directions.

enough to crush a submarine.

the pressure increases. [4]

1. a. How could you use a plastic bag, some water, and a sharp pin to demonstrate to another student that pressure acts equally in all directions in a liquid? Draw a diagram to illustrate your answer.

   ....

   ....

   ....

   .... [3]

   b. Describe how pressure varies with depth in liquids.

   ....

   .... [3]

   c. What is the pressure at the bottom of a swimming pool of depth 2.25 m if the density of water is 1000 kg/m³?

   pressure = ... [2]

Pressure

5.3 Pressure measurements

Language lab

Match the key words with their definitions.

manometer	an instrument used to measure atmospheric pressure
pressure	a push or a pull, measured in newtons
force	the force per unit area
barometer	an instrument used to measure the pressure due to a gas

[4]

1. a. Name an instrument that can be used to measure air pressure. .... [1]

   b. A student carries out an experiment using a drinks can. She pours 10 cm³ of water into the can and, carefully observing safety precautions, she heats the can on a tripod and gauze over a Bunsen flame. She allows the water to boil for about 30 s and then plunges the can into cold water to cool it. The drinks can immediately collapses. The results of the experiment are shown below.

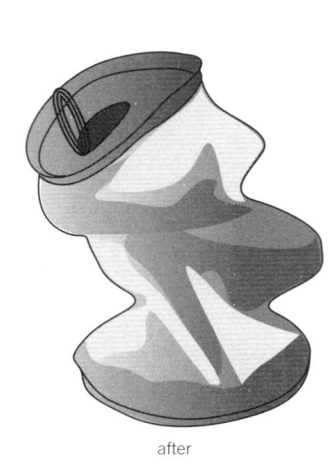

   before   after

   Explain why the drinks can was crushed.

   ....

   ....

   ....

   ....

   ....

   .... [6]

   c. Describe how a manometer could be used to find the pressure of the laboratory gas supply.

   ....

   ....

   .... [3]

Pressure — 5.4 Solids, liquids, and gases

Language lab

Use the key words to complete the sentences.

rows   melts   liquid   solid   boils   directions   arranged

In a .... the particles are arranged in .... and are close together. When a solid .... to form a .... the particles become randomly .... but are still close together. When a liquid .... to form a gas, the particles become much further apart and move randomly in all .... [7]

1. a. In the boxes, draw the arrangement of particles in a solid, liquid, and gas. [6]

   solid   liquid   gas

   b. Complete the statements for the states of matter.

solid	liquid	gas
	incompressible	
cannot flow		
		fills the entire container

   c. Explain why both solids and liquids are incompressible.

   ....
   .... [2]

   d. Explain why both liquids and gases can flow.

   ....
   .... [2]

   e. A student uses a microscope, a small glass box that can be opened and sealed shut, and a smoking piece of rope to study the movement of air particles.

   Describe and explain what the student will see when she observes the smoke cell under the microscope.

   ....
   ....
   ....
   ....
   .... [4]

Pressure

5.5 More about solids, liquids, and gases

Language lab

Complete the sentences using the key words.

| vibrating | negligible | around | apart | fixed | free | weaker | attraction |

In a solid the particles are .... in .... positions. There are strong forces of .... between particles and they are not .... to move .... In a liquid the forces of attraction between particles are .... than in a solid and so the particles are able to move around. In a gas the particles are very far .... and the forces of attraction between them are .... [8]

1. a. Describe the changes that occur in particle arrangement as ice melts and then evaporates.

   ....
   ....
   .... [3]

   b. Describe the changes that occur in particle movement as ice melts and then evaporates.

   ....
   ....
   .... [3]

2. Explain how a gas exerts a pressure on the walls of its container. In your explanation, make sure you use the keywords given below.

   | collisions | pressure | force | momentum |

   ....
   ....
   ....
   ....
   .... [4]

38

Pressure

5.6 Gas pressure and temperature

Language lab

Match the key words with their definitions.

velocity	a type of energy that raises the temperature of an object
heat	an atom or a molecule
pressure	a small glass box containing air and other particles
volume	distance travelled per second in a certain direction
particle	the force per unit area
smoke cell	the amount of space a substance takes up

[6]

1. a. By referring to the movement of the particles, explain the effect on pressure in a container if the gas in the container is heated and the container's volume does not change.

   ....
   ....
   ....
   ....
   .... [5]

   b. i. Which expand the most when heated: solids, liquids, or gases? .... [1]

   ii. Explain your answer by referring to the particle structure of solids, liquids, and gases.

   ....
   ....
   .... [2]

Pressure

5.7 Evaporation

> **Language lab**
>
> Use the key words below in your answer to **1. a.**
>
>   liquid   vapour   evaporating   particles   surface
>
> Use the key words below in your answer to **1. b.**
>
>   force   fastest   kinetic energy   escape   [9]

1. a. Look at the diagram and describe in as much detail as possible what is happening.

   ....

   ....

   .... [3]

   b. By referring to the forces between particles, explain which particles are most likely to leave the surface of the liquid.

   ....

   ....

   .... [3]

   c. When water vapour hits a cold surface, such as a mirror, condensation occurs. Describe and explain this process.

   ....

   ....

   .... [3]

2. A student has three thermometers and some cotton wool. Describe how he could use this apparatus, some water, and some acetone to investigate the effect of evaporation on temperature. Draw a labelled diagram of the apparatus.

   ....

   ....

   ....

   ....

   ....

   .... [5]

Pressure

5.8 Gas pressure and volume

Language lab

Match the key words with their definitions.

apparatus	close to the true value
accurate	an instrument used to measure pressure
pressure gauge	how one variable affects another
relationship	equipment for an experiment

[4]

1. a. A student uses the apparatus below to vary the pressure exerted on a fixed mass of dry air, and to measure the volume. Describe how she could do this as accurately as possible.

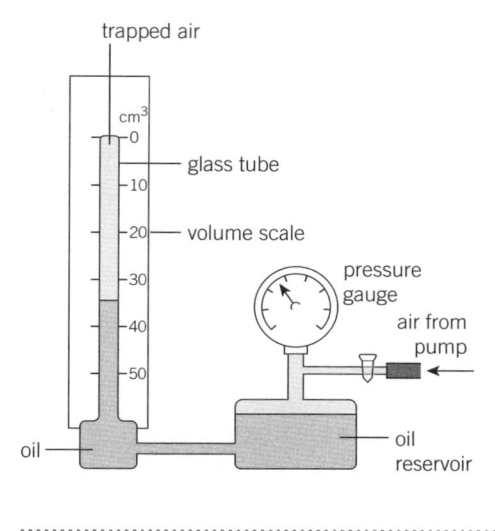

....
....
....
.... [4]

She obtains the following results.

pressure / kPa	100	150	200	250	300	350
volume / cm³	50	35	25	20	17	14

b. Plot a graph of her results. [4]

c. The student suggests that the volume is inversely proportional to the pressure. Use data from the graph to determine whether she is correct.

....
....
....
.... [3]

41

Pressure — Multiple choice questions

1. What is the pressure in Pa at the bottom of a column of mercury of height 736 mm if the density of mercury is 13 530 kg/m³?

   A 9.96×10^7
   B 9.96×10^3
   C 9.96×10^5
   D 9.96×10^4

2. What is the SI unit of pressure?

   A J
   B °C
   C s
   D Pa

3. Which of the following statements about pressure in gases is **incorrect**?

   A The particles collide with the walls of the container.
   B Decreasing the volume of the gas decreases the pressure.
   C The greater the temperature of the gas, the greater the speed of the particles.
   D The greater the temperature, the greater the force between the particles and the container.

4. Which of the following statements about liquids is **incorrect**?

   A Liquids cannot be compressed.
   B The particles can move past each other.
   C The particles are far apart.
   D Liquids can flow.

5. Which of the following statements about solids is **incorrect**?

   A Solids cannot be compressed.
   B The particles vibrate in fixed positions.
   C The particles can move past each other.
   D Solids cannot flow.

6. Which of the following statements about gases is **incorrect**?

   A Gases cannot be compressed.
   B The particles move randomly in all directions.
   C The particles are far apart.
   D Gases can flow.

7. Which of the following would **not** increase the rate of evaporation of water from a wet cloth?

   A Increase air flow across the cloth.
   B Increase the temperature of the surroundings.
   C Spread the cloth out.
   D Fold the cloth.

Thermal physics

6.1 Thermal expansion

Language lab

Anagrams: unjumble the key words needed in your explanation for question **1**.

rapture meet ....

rinses ace ....

ax spend ....

elk cub ....

[4]

1. The diagram below shows metal railway tracks on a cold day and a very hot day. Explain why there must be a gap between the sections of track.

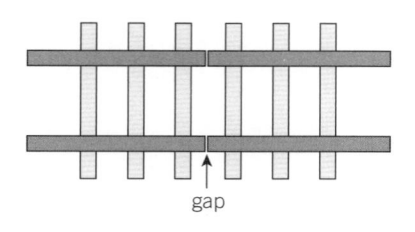

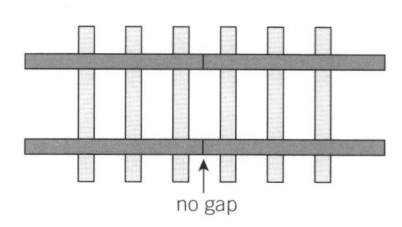

   ....
   ....
   ....
   .... [3]

Thermal physics 6.2 Thermometers

Language lab

Anagrams: unjumble the key words.

thirst more ....

cohere plum to ....

scurrying males ....

sluices .... [4]

1. a. Name three types of thermometer.

   .... [3]

   b. What are the fixed points on the Celsius scale of temperature?

   .... [1]

   c. On what property of liquids does a mercury-in-glass thermometer depend?

   .... [1]

   d. Give an advantage and a disadvantage of a data logger and temperature probe compared to using a mercury-in-glass thermometer.

   ....

   ....

   .... [2]

2. a. Complete the table comparing the Kelvin and Celsius scales of temperature.

	absolute zero	melting ice	boiling water
Kelvin scale	0 K		
Celsius scale	−273 °C		

   b. Explain what is meant by 'absolute zero'.

   ....

   ....

   .... [2]

   c. What is the effect on particle velocity if a gas is heated? Explain this.

   ....

   ....

   .... [3]

Thermal physics

6.3 More about thermometers

Language lab

Match the key words with their meanings.

capillary tube	a rubber stopper used to close a test tube
test tube	mark with a scale
boiling	turning from a liquid to a gas at constant temperature
bung	a thin glass tube closed at one end
calibrate	very thin glass tube [5]

1. A student fills a boiling tube with vegetable oil. She inserts a capillary tube through the bung in the top of the boiling tube, so that the oil is part way up the capillary tube. She places the boiling tube in some ice water at 0 °C and then in some boiling water at 100 °C.

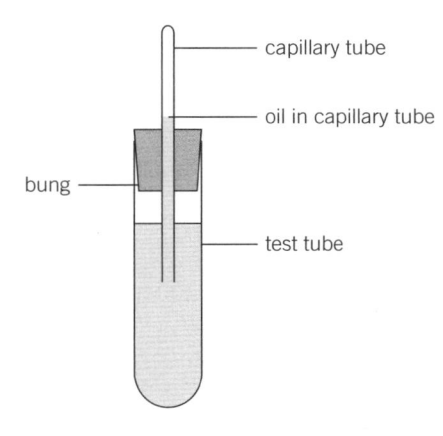

   a. Explain what would happen to the level of the oil in the capillary tube at 0 °C and at 100 °C.

   ....

   .... [2]

   b. Describe how she could calibrate her apparatus to allow her to find the temperature of some warm water.

   ....

   ....

   ....

   .... [4]

45

Thermal physics

6.4 Thermal capacity

Language lab

Anagrams: unjumble the key words.

cacti is a typeface chip ....

theorem term .... character tile eel .... [3]

1. a. Define the specific heat capacity of a material.

   .... [1]

   b. What is the unit of specific heat capacity? .... [1]

   c. i. 2400 J of energy is delivered to a metal block of mass 0.4 kg, raising its temperature by 15 °C. Find the specific heat capacity of the metal.

   specific heat capacity = ... [2]

   ii. How much energy is required to raise the temperature of 500 g of water by 10 °C if the specific heat capacity of water is 4200 J/kg °C?

   energy = ... [3]

   iii. 2 kJ of energy is delivered to a metal block of mass 200 g. What is the temperature rise of the block if the specific heat capacity of the metal is 350 J/kg °C?

   temperature rise = ... [3]

2. A student sets up the following apparatus in order to find by experiment the specific heat capacity of water.

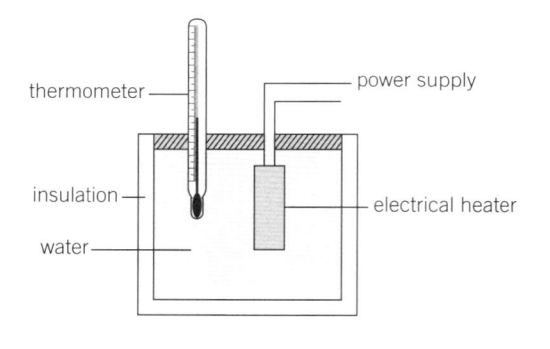

   (Diagram: thermometer, power supply, insulation, water, electrical heater)

   a. Describe how he should use the apparatus to find the specific heat capacity of water.

   ....

   .... [4]

   b. He obtains the following results:

   Starting temperature = 20 °C   End temperature = 42 °C

   Power supplied by the heater = 50 W   Heating time = 16 minutes

   Mass of water = 0.50 kg

   Use this data to calculate a value for the specific heat capacity of water.

   specific heat capacity = ... [4]

Thermal physics

6.5 Change of state

Language lab

Match the key words with their meanings.

beaker   not changing

Bunsen burner   a glass container used to contain liquids

constant   the degree of 'hotness', related to an object's internal energy

temperature
                                          a source of heat, which uses natural gas as a fuel   [4]

1. A student heats a block of wax in a beaker at a steady rate and records the temperature at regular time intervals. He keeps heating the wax until it is above its melting point.

   a. Sketch a graph of temperature against time for the wax. [3]

   b. i. Label the graph to show where the wax is solid, where it is liquid, and where it is melting. [3]

      ii. Explain the shape of the graph when the wax is solid.

      ....

      .... [2]

Thermal physics — 6.6 Specific latent heat

Language lab

Complete the sentences using the key words.

> thermometer   internal   mass   specific heat capacity   temperature

When a substance is heated, the .... energy increases and its ....

rises. The .... is the energy required to raise the temperature of 1 kilogram of the material

by 1 °C. The .... of material can be found using a balance and the temperature rise

measured with a .... [5]

1. a. How much energy is required to turn 0.45 kg of water to steam at the boiling point if the specific latent heat of vaporisation of water is 2.3 MJ/kg?

   energy = ... [2]

   b. Calculate the specific latent heat of fusion for ethanol if 65.4 kJ of energy are required to melt 0.6 kg of solid ethanol.

   specific latent heat of fusion = ... [2]

Thermal physics
6.7 Heat transfer (1): thermal conduction

Language lab

Match the key words with their meanings.

melt	does not easily pass on heat energy by particle vibrations
solidify	passes on heat energy by particle vibrations
conductor	an atom that has lost one or more electrons
insulator	a negatively charged particle that is not bound in an atom
free electron	change state from solid to liquid
ion	change state from liquid to solid [6]

1. A physics teacher sets up the following experiment. She attached metal drawing pins to a metal bar with melted wax, which then solidified, holding the pins in place. A Bunsen burner is placed at one end of the metal bar and the students observe the experiment over the next two minutes.

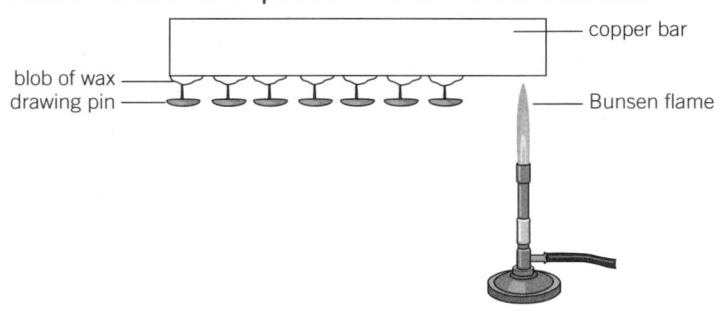

   [Diagram: copper bar with blobs of wax and drawing pins attached underneath, Bunsen flame at one end]

   a. Describe what happens to the drawing pins.
   .... [2]

   b. The metal bar is replaced by a wooden one. The end nearest the Bunsen burner begins to blacken and burn but the drawing pins do not fall off. Explain why.
   ....
   .... [2]

   c. What is an ion? .... [1]

   d. Describe how heat is transferred by conduction along the metal bar by the free electrons and ions.
   ....
   .... [2]

Thermal physics — 6.8 Heat transfer (2): convection

Language lab

Anagrams: unjumble the key words.

invent coco ....

deny its ....

rations pea ....

burns been run .... [4]

1. A physics teacher sets up an experiment to demonstrate convection. He drops some potassium permanganate crystals into one side of a beaker of cold water and then gently heats, with a Bunsen burner, under the potassium permanganate crystals.

   a. On the diagram, draw the path of warm water, which is indicated by shading.

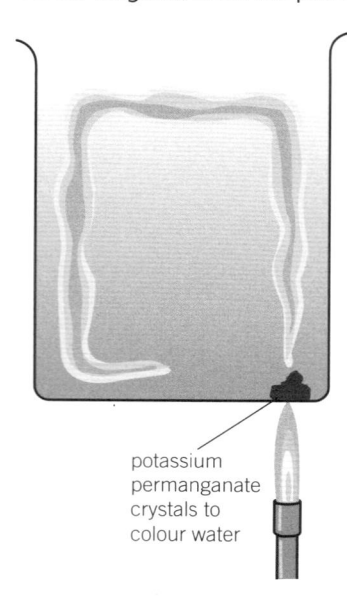

   potassium permanganate crystals to colour water

   b. Water that has been heated rises above cooler water in a beaker. By referring to density and the separation of the particles, explain why.

   ....

   .... [2]

   c. What happens when warm water cools? Explain this in terms of density and the separation of the particles.

   ....

   .... [2]

Thermal physics

6.9 Heat transfer (3): infrared radiation

Language lab

Match the key words with their meanings.

infrared radiation	heat that travels in electromagnetic waves
emission	taking in
absorption	bouncing back
reflection	giving out

[4]

1. A physics teacher sets up an experiment, as shown in the diagram, to investigate the effect of colour of surface on how much heat is emitted.

   She obtains the following results:

colour of surface	matt black	shiny black	silver	white
heat emitted (arbitrary units)	154	117	51	95

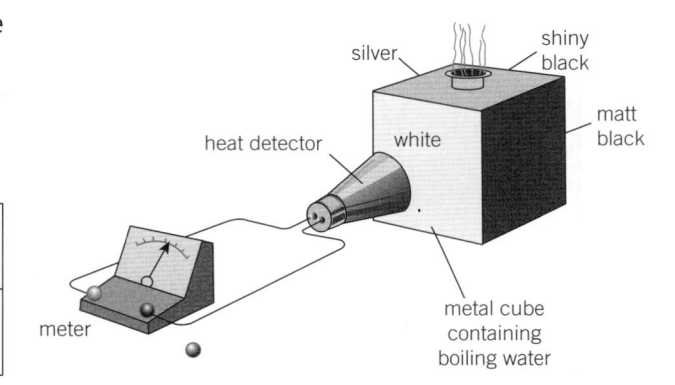

   a. Put the coloured surfaces in order from best to worst emitter of heat (infrared) radiation.

   .... [2]

   b. The student takes the initial reading of temperature from the two thermometers and switches on the lamp. She takes the temperature again after 15 minutes. Complete the results table.

colour of test tube	initial temperature / °C	final temperature / °C	temperature rise / °C
black	22	35	
silver	22	26	

   [2]

   c. Which is the better absorber of infrared radiation? Explain how you know this from the results.

   ....

   .... [2]

Thermal physics
6.10 Heat transfer at work

Language lab

Complete the word search.

Words may go downwards, upwards, forwards, backwards, or diagonally.

- absorb
- conduction
- convection
- density
- electrons
- fluid
- radiation
- reflect
- silver
- vacuum

A	Q	D	O	P	T	B	Y	W	L	G	D	Q	C	V
Y	C	L	O	J	X	D	O	A	B	S	O	R	B	L
A	F	A	B	R	A	D	I	A	T	I	O	N	L	I
X	A	B	T	B	L	T	C	E	L	F	E	R	F	I
N	K	Z	O	N	O	I	T	C	E	V	N	O	C	Y
F	B	B	M	P	I	S	E	G	N	H	N	R	P	R
Z	V	R	W	M	D	N	G	D	O	Z	R	I	T	E
V	N	Z	L	C	E	O	B	I	I	C	E	Z	M	Q
A	J	B	B	T	N	R	I	U	T	E	V	R	O	K
C	B	R	G	D	S	T	N	L	C	M	L	P	V	O
U	Y	W	N	B	I	C	J	F	U	K	I	T	T	Q
U	X	R	C	U	T	E	Y	D	D	T	S	S	C	D
M	Q	W	F	V	Y	L	E	P	N	A	N	L	H	D
G	O	V	F	Z	N	E	U	L	O	N	J	Z	M	L
Q	I	X	Q	L	O	C	B	G	C	W	V	X	D	X

[10]

1. The diagram below shows a vacuum flask, which can be used to keep drinks hot or cold. Explain how each of the labelled features of the flask reduces heat flow by conduction, convection, radiation, or evaporation.

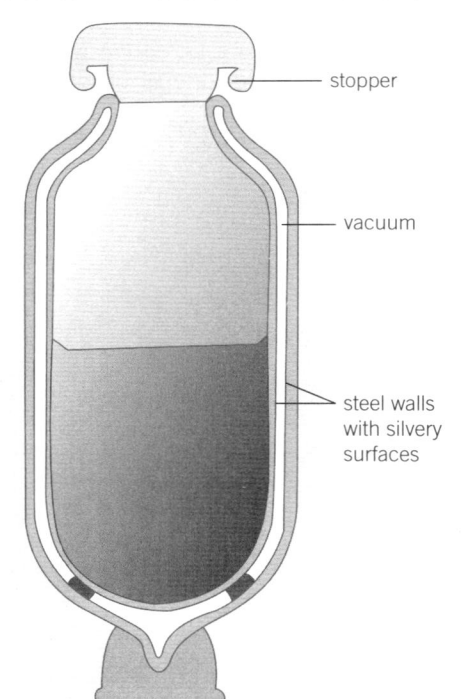

   - stopper
   - vacuum
   - steel walls with silvery surfaces

   A vacuum flask

   ....
   ....
   ....
   ....
   ....
   ....
   .... [6]

Thermal physics — Multiple choice questions

1. What is the SI unit of temperature?

   A J

   B °C

   C s

   D Pa

2. A liquid-in-glass thermometer depends on what property of a liquid?

   A Liquids expand on heating.

   B Pressure increases with temperature.

   C Liquids are incompressible.

   D Liquids can flow.

3. 1500 J of energy is supplied to a metal bar of mass 350 g. The temperature increases by 10 °C. What is the specific heat capacity of the metal?

   A 378 J/kg °C

   B 429 J/kg °C

   C 538 J/kg °C

   D 435 J/kg °C

4. 15 kJ of energy is supplied to 500 g of water. What is the increase in temperature if the specific heat capacity of water is 4200 J/kg °C?

   A 7 °C

   B 10 °C

   C 12 °C

   D 8 °C

5. How much energy is required to raise the temperature of 3.5 kg of steel, of specific heat capacity 450 J/kg °C, by 230 °C?

   A 0.4562 MJ

   B 314.32 kJ

   C 362.25 kJ

   D 1.389 MJ

6. Which of the following statements about heating a material is **incorrect**?

   A The greater the rate of supply of heat energy, the greater the rate of increase of temperature.

   B At the melting point, the temperature of the material stays constant.

   C The greater the heat capacity, the lower the rate of increase of temperature.

   D At the melting point, the kinetic energy of the particles increases.

7. How much energy is required to melt 250 g of copper, if its specific latent heat of fusion is 207 kJ/kg?

   A 0.4562 MJ

   B 314.32 kJ

   C 362.25 kJ

   D 51.75 kJ

8. Which of the following requires the greatest input of energy?

   A Changing 1 kg of water from liquid to gas at 100 °C (specific latent heat of vaporisation = 2.26 MJ/kg).

   B Increasing the temperature of 800 g of water by 35 °C (specific heat capacity = 4200 J/kg °C).

   C Changing 1 kg of water from solid to liquid at 0 °C (specific latent heat of fusion = 334 kJ/kg).

   D Increasing the temperature of 2 kg of copper by 250 °C (specific heat capacity = 385 J/kg °C).

Thermal physics — Multiple choice questions

9. Which of the following statements about evaporation and boiling is **incorrect**?

   A Boiling occurs at a fixed temperature.

   B Evaporation occurs over a range of temperatures.

   C Evaporation decreases the temperature of a liquid.

   D Boiling increases the temperature of a liquid.

10. Which of the following statements about heat flow is **correct**?

    A Materials containing trapped air are good insulators.

    B Black surfaces are poor emitters of infrared radiation.

    C Convection occurs in solids and liquids.

    D Water is a good conductor of heat.

Waves

7.1 Wave motion

Language lab

Complete the word search.

Words may go downwards, upwards, forwards, backwards, or diagonally.

- amplitude
- frequency
- hertz
- velocity
- vibration
- wave
- wavelength

Z	N	O	I	T	A	R	B	I	V	E	F	H	R	Z
A	A	G	Y	E	M	M	E	U	V	R	X	V	R	D
B	I	V	P	T	H	L	M	A	E	H	W	H	W	A
K	L	Y	E	Y	I	H	W	Q	J	V	V	M	K	J
Q	U	A	G	L	K	C	U	W	M	T	Q	F	Q	Q
X	W	M	N	S	E	E	O	X	O	M	E	A	D	L
J	E	P	I	U	N	N	R	L	N	V	C	O	R	L
N	Z	L	S	C	Q	B	G	B	E	H	K	M	B	X
D	B	I	Y	Z	J	J	Z	T	W	V	X	W	J	J
X	W	T	R	T	U	R	M	I	H	M	T	B	D	T
K	B	U	O	R	N	D	A	B	O	Z	K	T	R	A
B	N	D	G	E	C	S	B	H	D	L	T	C	I	K
M	H	E	S	H	G	W	J	C	Q	S	E	X	U	W
U	S	O	Z	Z	K	T	A	I	H	L	W	E	D	B
T	T	Z	B	L	V	J	H	Q	L	Y	O	O	C	C

[7]

1. What type of wave is a sound wave?... [1]

2. a. i. What is the definition of frequency of a wave?

.... [1]

ii. What is the unit of frequency? [1]

b. Write down the equation that links speed, wavelength, and frequency of a wave. [1]

c. What is the speed of a wave of wavelength 2.3 m and frequency 100 Hz?

speed = ... [2]

55

Waves
7.2 Transverse and longitudinal waves

Language lab

Anagrams: unjumble the key words.

vat gel when ....

ravens rest ....

tedium pal ....

diluting loan ....

bravo in it ....

fern arts ....

piccolo loses ....

vacate due .... [8]

1. a. A student vibrates the end of a long spring to create a wave. Firstly, he moves his hand back and forth along the spring. Then he moves his hand from side to side at right angles to the spring, as shown in the diagrams.

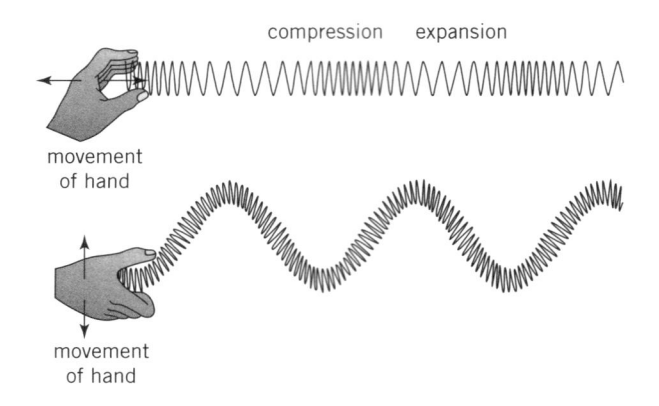

   Describe what happens to the coils of the spring when the wave travels from his hand.

   ....

   .... [4]

   b. i. What is the name given to the distance between adjacent particles along a wave at the same point in their vibration? Mark this onto both of the diagrams in part **a**. [1]

   ii. What is the name given to the distance between the maximum displacement in a wave (crest) and the equilibrium (undisplaced) position? Mark this onto the bottom diagram in part **a**. [1]

Waves

7.3 Wave properties (1): reflection and refraction

Language lab

Complete the sentences using the key words.

| unchanged | deep | decreases | slows | wavefronts |

As a water wave travels from .... to shallower water, it .... down and its wavelength .... The .... get closer together. The frequency of the wave is .... [5]

1. a. A student wishes to model the behaviour of light as it moves from one medium to another using water waves. He sets up a water tank and, by dipping a metal rod into the water, creates a wave which travels from an area of deep water to an area of shallow water. The speed of the wave decreases in the shallow region. The diagram (right) shows a view of the water tank from above with a series of wavefronts arriving at the boundary between the deep and the shallow water.

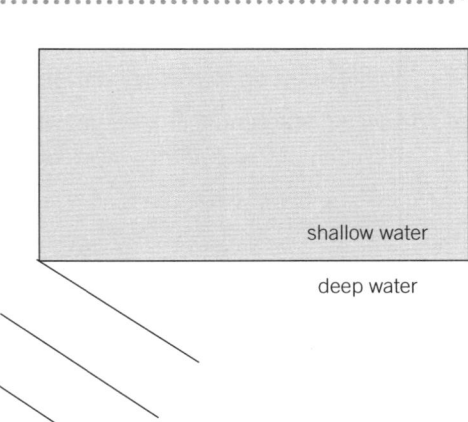

    i. Complete the diagram to show the wavefronts after the wave enters the shallow water. [2]

    ii. On the diagram, mark the direction of travel of the wave in both deep and shallow water. [2]

    iii. Explain why you have drawn the wavefronts in this position.

    ....
    ....
    .... [3]

   b. What is the angle between a wavefront and the direction of travel of a wave? .... [1]

   c. The speed of a water wave decreases as the wave enters shallower water. What happens to the wavelength of the wave?

    .... [1]

2. a. i. What is the range of frequencies that a human ear can hear?

    .... [1]

    ii. What is the name given to sounds with frequencies above this range?

    .... [1]

Waves

7.4 Wave properties (2): diffraction

Language lab

Complete the crossword.

Across

1. The type of wave where the oscillations are perpendicular to the direction of energy transfer
3. Used to demonstrate the behaviour of water waves
6. An area within a longitudinal wave where the particles are closest together
8. The unit of wavelength
10. The type of wave where the oscillations are parallel to the direction of energy transfer
13. Can be represented as lines perpendicular to the direction of travel of the wave
15. The number of waves passing a point per second

Down

2. An area of a longitudinal wave where the particles are furthest apart
4. The unit of frequency
5. The distance between adjacent points on a wave that are at the same point in their oscillation
7. An example of a longitudinal wave
9. An example of a transverse wave
11. The distance from the centre of a vibration to the maximum displacement
12. The distance travelled per second
14. Waves transfer this without transferring matter

[15]

1. A water wave approaches a gap in a barrier. Complete the diagrams to show what would happen to the wave if the gap was smaller than the wavelength, approximately equal to the wavelength, and much larger than the wavelength.

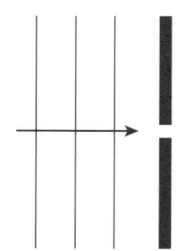

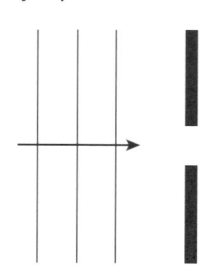

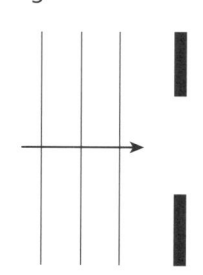

   What is the name for this effect? .... [1]

58

Waves — Multiple choice questions

1. What is the unit of wavelength?

    A J
    B m
    C s
    D km

2. What is the unit of frequency?

    A kg
    B °C
    C s
    D Hz

3. What is the velocity of a sound wave of frequency 20 kHz and wavelength 75 mm?

    A 150 000 mm/s
    B 267 m/s
    C 1500 m/s
    D 15 mm/s

4. What is the angle between a wavefront and the direction of travel of a wave?

    A 90°
    B 45°
    C 0°
    D 180°

5. What is the effect on a water wave of entering shallower water?

    A Frequency increases
    B Frequency decreases
    C Velocity decreases
    D Velocity increases

6. Which of the following statements about diffraction is **incorrect**?

    A If the wavelength of the wave is larger than the gap, the wave is reflected.

    B If the wavelength of the wave is much larger than the gap, the wave is diffracted through the gap.

    C Maximum diffraction occurs when the wavelength and the gap are approximately equal.

    D If the gap is much larger than the wavelength, there is very little diffraction.

7. A water wave takes 6.2 s to travel from a boat to the harbour wall and back. The total distance to the wall and back is 5.0 m. What is the speed of the wave?

    A 0.81 m/s
    B 1.6 m/s
    C 31 m/s
    D 15.5 m/s

Light

8.1 Reflection of light

Language lab

Match the key words with their meanings.

plane mirror	an image that cannot be focused on a screen
normal	angle between the incident ray and the normal
angle of incidence	angle between the reflected ray and the normal
angle of reflection	an image that can be focused on a screen
real	a narrow beam of light
virtual	a constructed line at right angles to a surface
ray	a flat reflective surface [7]

1. A teacher demonstrates laser light to a group of students. She puts the laser on the floor and allows the beam to hit the wall at the far side of the laboratory. The students observe a bright red spot on the wall.

    a. The students all stand up during this experiment. Why is this important?

    .... [1]

    b. The students are unable to see the laser beam until the teacher dusts some fine powder into the beam. Explain why.

    ....
    ....
    .... [2]

2. a. Describe the image formed in a plane mirror.

    ....
    .... [3]

    b. What is the difference between a virtual and a real image?

    .... [1]

Light

8.2 Refraction of light

Language lab

Complete the sentences using the key words.

| away | normal | refraction | speeds | slows | density |

When light travels from air to glass, it .... down and bends towards the ....

This is due to the change in .... of the material. When light travels from glass to air,

it .... up and bends .... from the normal. This phenomenon is called

.... [6]

1. a. What is meant by refraction?

   ....

   .... [1]

   b. When a light ray moves from air into glass, what happens to the speed of light?

   .... [1]

   c. Complete the diagram to show the light ray entering and leaving the glass block. Include reflected as well as refracted rays.

   [4]

2. White light is incident on a triangular glass prism as shown. Complete the path of the ray through the prism and label the diagram.

   [3]

   a. Explain why light emerges from the prism in this way.

   ....

   ....

   .... [3]

   b. What is the name of this effect? .... [1]

61

Light — 8.3 Refractive index

Language lab

Anagrams: unjumble the key words.

conifer art .... exacted river fin ....

patterns ran .... ghosted elf pi .... [4]

1. a. Write down the equation relating angle of incidence to angle of refraction. [1]

   b. i. The angle of incidence for a light ray entering a transparent plastic block is 45° and the angle of refraction inside the block is 30°. Calculate the refractive index for the plastic.

   refractive index = ... [2]

   ii. A ray of light is incident in air on a glass block at 40°. What is the angle of refraction if the refractive index is 1.4?

   angle of refraction = ... [3]

   iii. Calculate the angle of incidence for a ray of light if the angle of refraction in a plastic block of refractive index 1.6 is 25°.

   angle of incidence = ... [3]

   c. For the diagram (right) use a protractor to measure the angle of incidence and angle of reflection. Use your values to calculate the refractive index of the glass block.

   refractive index = ... [4]

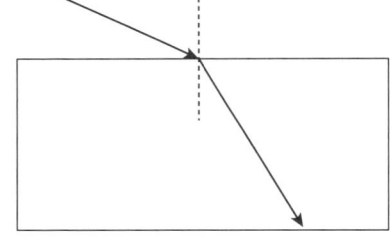

   d. i. Write down the formula for calculating the critical angle for a glass–air boundary. [1]

   ii. What is the refractive index of a material if the critical angle is 55°?

   refractive index = ... [2]

   iii. Calculate the critical angle for light travelling from glass to air if the refractive index of the glass is 1.5.

   critical angle = ... [3]

Light

8.4 Total internal reflection

Language lab

Match the key words with their meanings.

refracted ray	when all light rays are reflected inside a more dense medium
critical angle	beam of light that has changed direction on entering a different medium
total internal reflection	very thin glass rod down which light travels by total internal reflection
optical fibre	where one medium ends and another begins
boundary	the incident angle beyond which light is totally internally reflected [5]

1. a. The diagram below shows the incident and refracted rays for a ray incident inside a glass block. Describe what would happen to the refracted ray if the angle of incidence is increased.

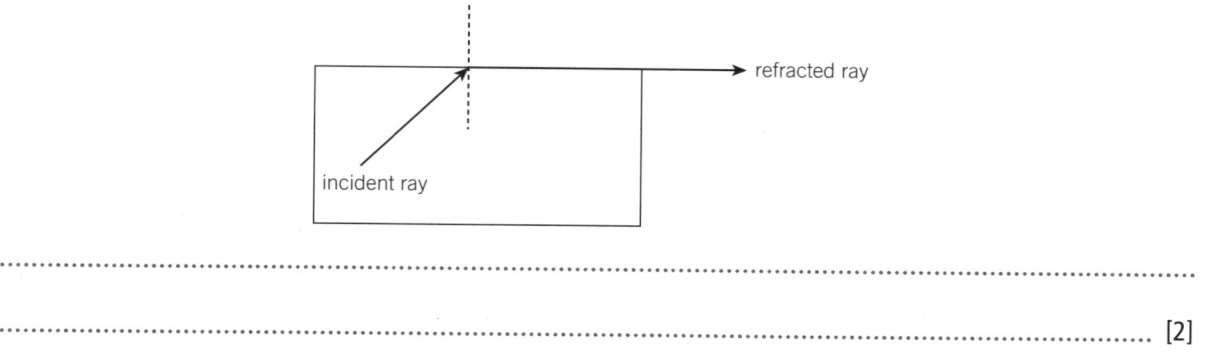

    ....

    .... [2]

   b. i. On the diagram below, complete the path of the light ray, which totally internally reflects along an optical fibre. Mark on the core and the cladding.

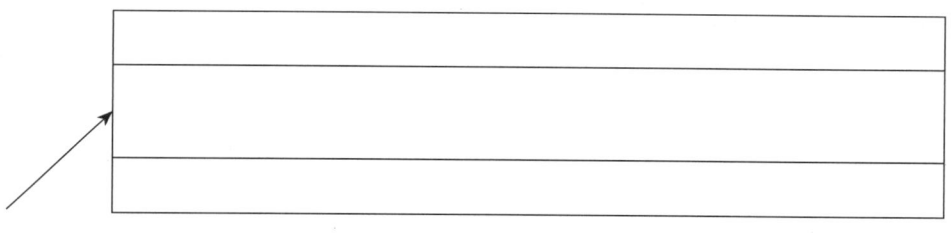

      [5]

      ii. In an endoscope, which is used in medicine to look inside the body and perform procedures such as biopsies, there are many optical fibres similar to the one shown in c. i. Explain how an endoscope can be used to image inside the body.

      ....

      ....

      .... [3]

      iii. Explain why many fine optical fibres, all lying parallel to each other, produce a better quality image than a few thicker fibres.

      ....

      ....

      ....

      .... [3]

Light

8.5 The converging lens

Language lab

Complete the sentences using the key words.

| focal length | principal axis | rays | principal focus | convex | lens |

When light .... which were originally parallel and close to the ....

pass through a .... converging .... , they converge to a point called the

.... The distance between the centre of the lens and the principal focus is called the

....

[6]

1. a. Mark the focal length and principal focus of the converging and diverging lenses on the diagram.

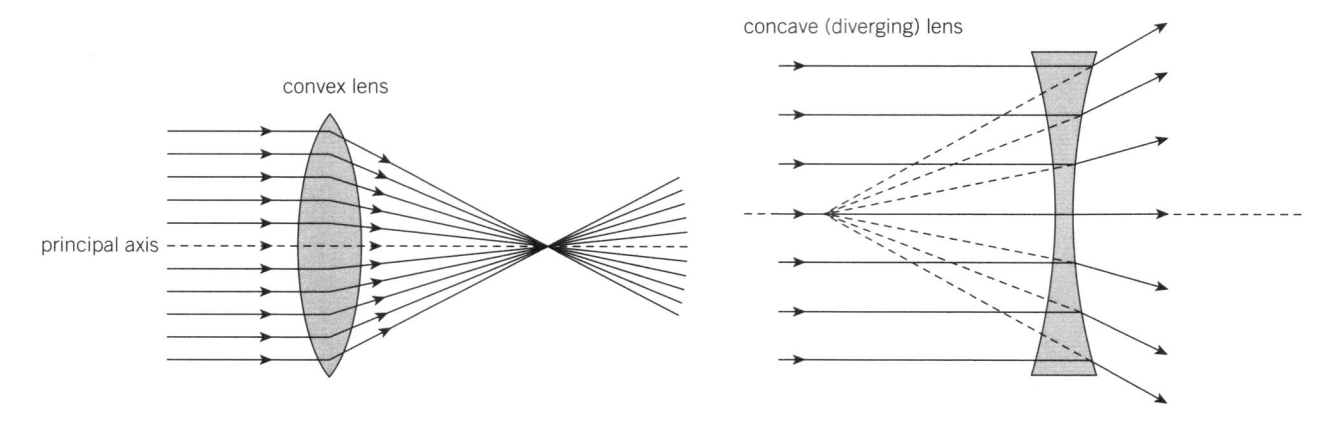

   convex lens

   concave (diverging) lens

   principal axis

   b. i. Complete the ray diagrams to show the path of two rays from the object through the convex lens.

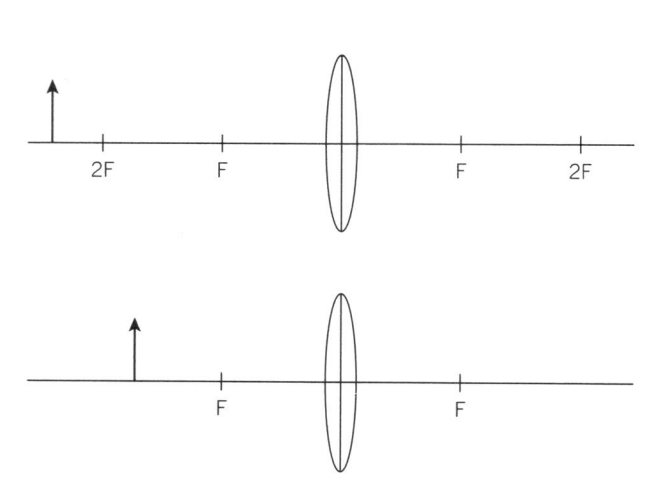

   [4]

   ii. Mark on the image and state the nature of the image in each case. [4]

   c. Suggest a use for the convex lens in each of the arrangements in **b**.

   i. ....

   ii. .... [2]

Light — 8.6 Applications of the converging lens

Language lab

Anagrams: unjumble the key words.

hi mind side ....

feign maid ....

earl ....

hug trip ....

rut vial ....

river tend .... [6]

1. a. i. On the grid below, draw the principal axis and a convex lens. The lens has a focal length of 20 cm. Choose an appropriate scale and mark on the focal point before and after the lens. [2]

   ii. The lens is to be used as a magnifying glass. Draw an object in the appropriate position. [1]

   iii. Complete the diagram by drawing rays from the object to locate the image and draw on the eye of the observer. [4]

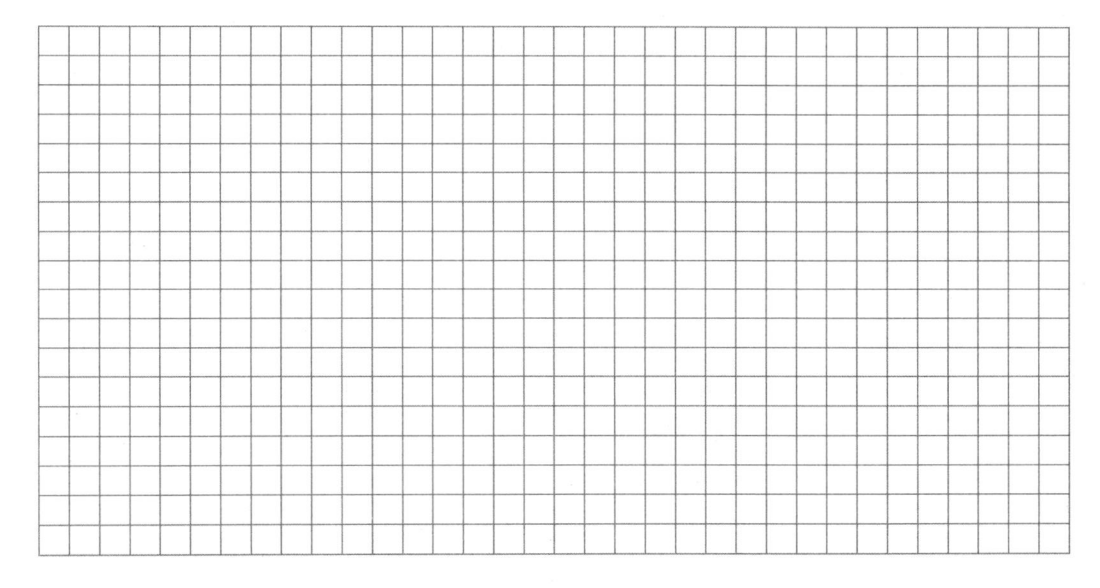

   b. State the nature of the image. [3]

   ....

Light

8.7 Electromagnetic waves

Language lab

Anagrams: unjumble the key words.

voice warm ....

rivulet alto ....

friar den ....

or aid .... [4]

1. a. Name three characteristics that all electromagnetic waves have in common.

   ....

   ....

   .... [3]

   b. Write a mnemonic to help you remember the order of the electromagnetic spectrum.

   ....

   ....

   .... [1]

   c. i. Which section of the electromagnetic spectrum has the highest frequency radiation?

   .... [1]

   ii. Which section of the electromagnetic spectrum has the longest wavelength radiation?

   .... [1]

   iii. Name three types of electromagnetic radiation that can be used for communication.

   .... [3]

   iv. Name three types of electromagnetic radiation that are ionising.

   .... [3]

   d. i. Name a use of X-rays.

   .... [1]

   ii. Which type of electromagnetic radiation is released from an unstable nucleus?

   .... [1]

   iii. Which type of electromagnetic radiation can cause sunburn?

   .... [1]

   iv. Which type of electromagnetic radiation is used to rapidly heat food?

   .... [1]

   e. What is the frequency of a wave of speed 300 million m/s and wavelength 100 m?

   frequency = ... [2]

Light

8.8 Applications of electromagnetic waves

Language lab

Complete the word search.

Words may go downwards, upwards, forwards, backwards, or diagonally.

- communication
- electromagnetic
- gamma
- image
- ionising
- microwave
- radio
- signal
- sunbed
- visible

E	N	I	Z	T	O	F	G	Q	E	D	O	B	Y	H
I	V	F	D	P	X	E	N	N	L	P	K	B	H	T
R	G	A	A	T	L	I	I	Q	E	C	B	H	L	T
I	T	J	W	B	M	X	S	G	C	R	J	W	O	D
S	A	D	I	O	N	I	I	L	T	C	D	G	L	Z
B	I	S	E	B	R	H	N	A	R	S	H	M	B	B
F	I	G	Y	H	T	C	O	E	O	D	P	I	I	E
V	B	M	N	M	J	A	I	I	M	X	G	B	D	U
W	O	Z	T	A	B	D	W	M	A	E	G	A	M	I
K	I	T	D	R	L	G	R	R	G	R	Y	U	V	Y
E	Z	N	O	I	T	A	C	I	N	U	M	M	O	C
S	U	N	B	E	D	M	S	C	E	M	D	F	V	P
X	A	L	X	I	W	M	X	Z	T	H	C	B	X	H
A	B	E	O	F	F	A	L	D	I	A	X	H	C	W
I	K	X	I	O	D	U	T	M	C	H	P	Q	D	L

[10]

1. a. Explain how X-rays can be used to obtain an image of a broken bone.

   ....

   ....

   .... [3]

   b. Explain the dangers of sunbeds to someone who is considering using one.

   ....

   ....

   .... [3]

   c. Explain how night vision cameras can be used by the police to track criminals at night.

   ....

   ....

   .... [3]

Light — Multiple choice questions

1. What is the frequency of a radio wave of wavelength 1.5 km and velocity 300 million m/s?

    A   200 kHz
    B   20 kHz
    C   450 kHz
    D   450 MHz

2. What is the wavelength of microwave radiation of velocity 300 million m/s and frequency 10 GHz (1 GHz = 1 000 000 000 Hz)

    A   3.0 m
    B   0.03 m
    C   33 m
    D   33 cm

3. Which of the following statements about an image in a plane mirror is **incorrect**?

    A   The image is laterally inverted.
    B   The image is the same distance behind the mirror as the object is in front.
    C   The image is real.
    D   The image is the same size as the object.

4. Which of the following statements about the reflection of light is **incorrect**?

    A   The angle of incidence is equal to the angle of reflection.
    B   The normal is perpendicular to the mirror surface.
    C   Increasing the angle of incidence increases the angle of reflection.
    D   The normal, incident ray, and reflected ray lie in different planes.

5. What is the refractive index of plastic if it slows the speed of light from 300 million m/s to 240 million m/s?

    A   1.25 million
    B   0.8 million
    C   1.25
    D   0.8

6. What is the angle of refraction in a glass block of refractive index 1.52 if the angle of incidence is 32°?

    A   20°
    B   54°
    C   23°
    D   60°

7. Light is incident in air on a glass surface at the critical angle. What is the angle of refraction?

    A   90°
    B   45°
    C   0°
    D   180°

8. What is the critical angle for plastic of refractive index 1.42?

    A   55°
    B   45°
    C   65°
    D   35°

Light — Multiple choice questions

9. Light is incident in air on a glass surface at 90°. What is the angle of refraction?

   A 90°
   B 45°
   C 0°
   D 180°

10. When light passes through a triangular glass prism, which colour is deviated through the greatest angle?

    A red
    B violet
    C green
    D yellow

11. Which colour light travels fastest in glass?

    A red
    B violet
    C green
    D yellow

12. What is the name of the process by which white light spreads out into the colours of the spectrum after passing through a glass prism?

    A diffraction
    B refraction
    C dispersion
    D reflection

13. What is the name of the process by which light changes direction as it passes from air to glass?

    A diffraction
    B refraction
    C dispersion
    D reflection

14. What type of image is produced when a converging lens is used as a magnifying glass?

    A diminished
    B inverted
    C magnified
    D real

15. What type of image is produced when a converging lens is used to focus light in a camera?

    A diminished
    B upright
    C magnified
    D virtual

16. What type of electromagnetic radiation is emitted from hot objects?

    A infrared
    B microwave
    C radio
    D ultraviolet

17. What type of electromagnetic radiation is used in security marking valuables?

    A infrared
    B microwave
    C radio
    D ultraviolet

Sound

9.1 Sound waves

Language lab

Match the key words with their meanings.

tuning fork   not containing any particles

vibration   a vibrating metal instrument that produces a sound of constant frequency

evacuated   back and forth motion   [3]

1. a. A student strikes the tuning fork, shown below, by hitting it against a piece of cork. He observes the motion of the two prongs of the tuning fork and holds it to his ear.

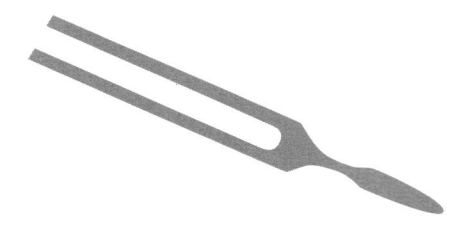

   Describe what he observes and what he hears.

   ....

   ....

   .... [3]

   b. Explain what he would observe if the vibrating tuning fork were placed in an evacuated glass jar.

   ....

   .... [2]

2. A boat uses sonar to survey the ocean floor. A high-frequency sound wave is reflected from the ocean floor and received 0.20 s later. The speed of sound in water is 1500 m/s.

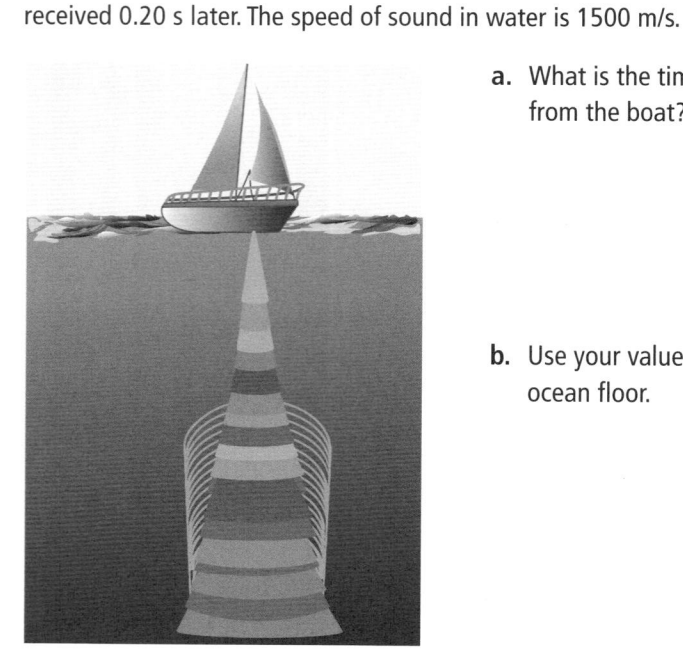

   a. What is the time taken for the sound pulse to reach the ocean floor from the boat?

      time = ... [1]

   b. Use your value to calculate the distance between the boat and the ocean floor.

      distance = ... [2]

70

Sound

9.2 Properties of sound

Language lab

Anagrams: unjumble the key words.

crimson pose ....

fraction ear ....

collies scoop ....

tulip made .... [4]

1. a. i. A sound wave is formed from a series of compressions and rarefactions. Draw a diagram of a sound wave, labelling the compressions and rarefactions.

   [3]

   ii. Describe the variation in air pressure in a sound wave in moving from a compression to a rarefaction.

   ....
   ....
   .... [2]

   iii. What is the wavelength of a wave of speed 330 m/s with a frequency of 2 kHz?

   wavelength = ... [2]

   b. i. A sound wave is a longitudinal wave but it can be represented on an oscilloscope screen as shown.

   On the grid, sketch a waveform that represents a sound wave of half the frequency and a third of the amplitude. [2]

   ii. What would an observer hear when these changes were made to the sound wave?

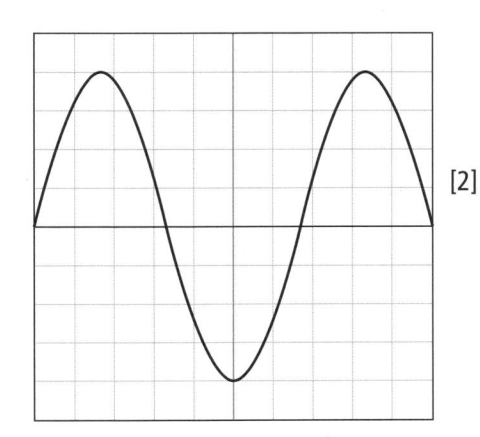

   ....
   .... [2]

71

Sound

9.3 The speed of sound

Language lab

Complete the sentences with the missing words to describe what is happening in the diagram shown in question **1. a.**

| time | microphone B | microphone A | stops | starts | speed | wave |

When the hammer strikes the metal block a sound .... travels to .... and

.... the timer. When the sound reaches ...., the timer ....

The student can calculate the .... by dividing the distance between the two microphones by the

.... on the timer. [7]

1. a. A student uses the apparatus shown to find by experiment the speed of sound in air.

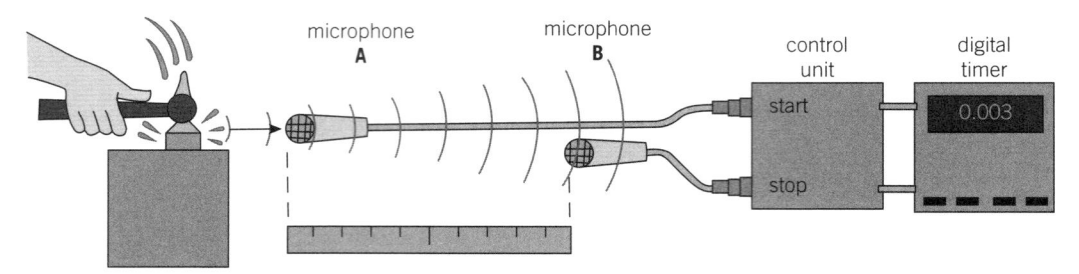

   She strikes the metal plate with the hammer. Describe how microphone **A** and microphone **B** can be used to measure the time for the sound wave to travel the distance indicated by the metre rule.

   She obtains the following results from the experiment:

distance travelled / m	time /
1.00	0.0030
1.00	0.0031
1.00	0.0029
1.00	0.0032
1.00	0.0030

   i. Complete the table to include the unit for time. [1]

   ii. Why did the teacher record the distance as 1.00 in the table instead of 1?

   .... [1]

   iii. Why did the teacher take so many repeat readings of time?

   .... [1]

   iv. Calculate the average time.

   average time = ... [1]

   v. Use the average time to calculate the speed of sound in air and give the correct unit.

   speed of sound = ... [2]

   b. A girl claps her hands and hears the echo from a distant wall 2 s later. Sound travels at 330 m/s in air. Find the distance between the girl and the wall.

   distance = ... [3]

Sound

9.4 Musical sounds

Language lab

Match the words with their meanings.

noise	used to display the waveforms of sound waves
vocal cord	an instrument that produces a sound when struck
percussion	sound waves that vary randomly in frequency and amplitude
amplitude	the vibrating membranes that produce sound in humans
pitch	the maximum displacement of a particle from the equilibrium position
oscilloscope	used to produce waves of varying amplitude and frequency
signal generator	the frequency of the sound—that is, how high the note sounds [7]

1. a. A student uses a signal generator and loudspeaker to vary the amplitude of a sound wave. She listens to the sound and observes the waveform on an oscilloscope screen.

   Describe and explain the relationship between the height of the waveform on the screen, the amplitude of the wave produced by the signal generator, and the sound she hears.

   ....

   ....

   .... [3]

   b. The student now uses the signal generator to vary the frequency of the wave. Describe and explain the relationship between the number of waves she sees on the screen, the frequency of the wave produced by the signal generator, and the sound she hears.

   ....

   ....

   .... [4]

2. a. A student uses a microphone and an oscilloscope to display waveforms from different instruments producing sound of the same pitch. She sings a note, plucks a string, and blows into a hollow tube. In each case describe how the sound is made.

   voice ....

   .... [1]

   string ....

   .... [1]

   tube ....

   .... [1]

   b. How would the waveforms in **a.** differ? .... [1]

Sound

Multiple choice questions

1. Which of the following is a longitudinal wave?

   A radio waves
   B microwaves
   C water waves
   D sound in air

2. Which of the following increases the volume of sound when increased in magnitude?

   A frequency
   B wavelength
   C speed
   D amplitude

3. Which of the following types of wave have vibrations parallel to the direction of energy transfer?

   A water waves
   B sound in air
   C radio waves
   D ultrasound waves

4. Which of the following statements about pressure in sound waves travelling in air is **correct**?

   A The air particles are closest together in compressions.
   B The pressure is highest in a rarefaction.
   C The air particles are closest together in a rarefaction.
   D The distance between adjacent compressions is equal to half a wavelength.

5. In which material does sound travel fastest?

   A water
   B air
   C metal
   D wood

Magnetism

10.1 Magnets

Language lab

Anagrams: unjumble the key words.

gnat me ....

pelt honor ....

house plot ....

fraction merge .... [4]

1. a. i. What is the effect of a north magnetic pole on a south pole?

   .... [1]

   ii. What is the effect of a south magnetic pole on another south pole?

   .... [1]

   b. What is the most common ferromagnetic element? .... [1]

2. A student has three objects, **A**, **B**, **C**, and a bar magnet.

   Both ends of object **A** attract to the south pole of the magnet. Object **B** does not attract the south pole at all. Object **C** attracts the south pole at one end and repels at the other end.

   Suggest what each of the bars is made from.

   A ....

   B ....

   C .... [3]

75

Magnetism

10.2 Magnetic fields

Language lab

Match the words with their meanings.

attract	a ferromagnetic material in which the domains are not lined up
repel	a tiny magnet found inside a magnetic material
magnet	a particular type of magnetism most commonly found in iron
ferromagnetic	pull towards with a force
domain	a magnetised piece of iron
unmagnetised	push away with a force
magnetic field	the area around a magnet where another magnet experiences a force

[7]

1. a. Describe one way to magnetise a piece of iron.

   .... [1]

   b. Describe how a piece of iron can be demagnetised.

   .... [1]

2. a. What is the name for the area around a magnet inside which another magnet experiences a force?

   .... [1]

   b. Draw the field lines around the bar magnet. [3]

   | N   S |

   c. What do the arrows on field lines represent?

   .... [1]

   d. Why are the field lines closer together near the poles of the magnet?

   .... [1]

3. Draw the field lines for each of the following pairs of magnets. [6]

   a. [N]   [S]   b. [N]   [N]

76

Magnetism

10.3 More about magnetic materials

Language lab

Complete the word search.

Words may go downwards, upwards, forwards, backwards, or diagonally.

- attract
- domain
- ferromagnetic
- field
- magnet
- north
- repel
- south
- unmagnetised [9]

P	H	C	P	A	G	C	Y	F	H	F	X	U	P	J
O	R	Q	Q	U	T	I	C	M	D	T	P	X	J	Q
B	D	E	Z	U	X	T	F	X	X	Z	R	E	Z	D
C	J	S	H	L	C	E	R	J	A	T	T	O	Q	B
N	D	P	Z	E	H	N	H	A	Z	J	U	B	N	F
M	M	B	E	P	B	G	N	A	C	K	G	R	N	C
V	A	P	X	E	D	A	Q	H	Z	T	K	H	Y	J
K	D	H	Z	R	O	M	Y	M	T	M	M	Z	Y	R
X	E	G	E	C	M	O	T	H	A	E	H	X	Z	X
W	F	C	M	M	A	R	N	G	U	V	L	W	N	W
E	S	H	T	B	I	R	N	P	E	R	Q	U	Y	D
U	N	M	A	G	N	E	T	I	S	E	D	S	L	W
I	Z	X	Q	H	T	F	H	T	U	O	S	E	P	S
Q	J	J	P	M	N	Z	Z	B	A	H	I	Y	W	G
O	K	B	E	T	L	N	M	X	Z	F	D	G	E	R

1. A student makes the following prediction:

   'The higher the current through an electromagnet, the stronger the magnetic field.'

   She builds the circuit shown (right).

   a. Describe how she could use this circuit and several small paper clips to test her prediction.

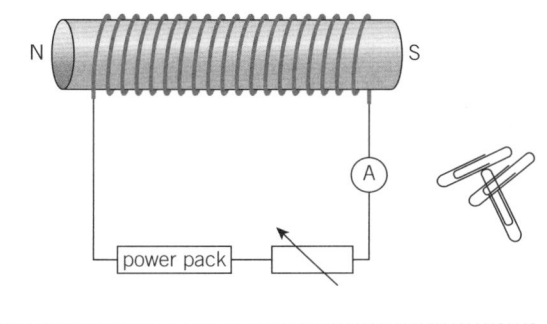

   ....

   ....

   ....

   ....

   ....

   .... [4]

The student obtains the following results:

current / A	0.2	0.4	0.6	0.8	1.0	1.2	1.4	1.6	1.8	2.0
number of paper clips held	1	2	2	4	5	7	7	8	9	10

77

Magnetism — 10.3 (continued)

b. Plot a graph of number of paper clips against current and draw a line of best fit. [4]

c. What is the effect on the number of paper clips held by the electromagnet if the current is doubled?

.... [1]

d. Use the table and graph to predict the number of paper clips held if the current was increased to 2.6 A.

.... [1]

Magnetism — Multiple choice questions

1. What are the tiny magnets inside a ferromagnetic material called?

   A   mini magnets
   B   grains
   C   domains
   D   cells

2. What is the name for the temporary magnetism when a magnet picks up a piece of iron?

   A   induced
   B   repulsion
   C   momentary
   D   weak

3. Which way do the arrows on the magnetic field lines around a bar magnet point?

   A   To the north pole of the magnet
   B   To the north pole of the Earth
   C   To the south pole of the magnet
   D   To the south pole of the Earth

4. The most easily magnetised and demagnetised material is:

   A   steel
   B   aluminium
   C   copper
   D   iron

Electric charge

11.1 Static electricity

Language lab

Match the key words with their meanings.

charged	contains charged particles that are not free to move
coulombmeter	the unit of charge
electrostatic	containing an unbalanced number of positive and negative charges
insulator	used to measure charge
coulomb	the type of force between charged particles

[5]

1. a. A group of IGCSE students were experimenting with charging. They had rods made of two different insulating materials, polythene and acetate, which they rubbed with cotton and silk cloths. They then used a coulombmeter to test the charge on the rods.

   After rubbing the polythene rod with the cotton cloth, they discovered the rod had a negative charge. After rubbing the acetate rod with the silk cloth, they discovered the rod had a positive charge.

   i. On the diagrams below, indicate the charges on the rods and the cloths.

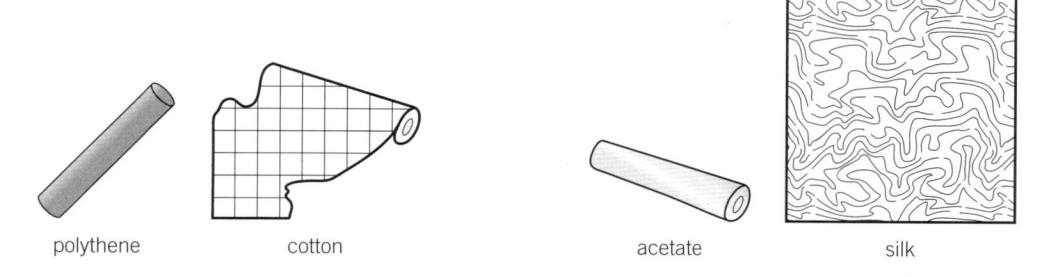

   polythene   cotton   acetate   silk

   ii. Which charged particles move in the charging process? .... [1]

   iii. Which force causes charging? .... [1]

   iv. What is the unit of charge? .... [1]

   v. Draw an arrow on each diagram to indicate the direction that the charged particles have moved. [2]

   b. Using a piece of string, the students suspend the charged polythene rod from a clamp stand so that it is free to move and bring the charged acetate rod towards it.

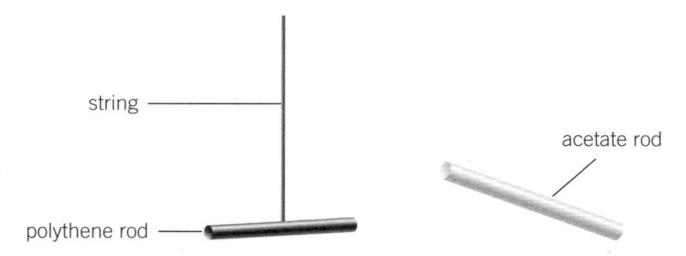

   string
   acetate rod
   polythene rod

   i. What is the effect of the charged acetate rod on the polythene rod?

   .... [1]

   ii. What would be the effect if two charged polythene rods were used instead?

   .... [1]

79

Electric charge

11.2 Electric fields

Language lab

Complete the word search.

Words may go downwards, upwards, forwards, backwards, or diagonally.

- charge
- coulombmeter
- electron
- electrostatic
- field
- insulator
- ion
- negative
- positive

[9]

C	V	P	X	M	X	D	E	O	R	K	C	R	S	Q
R	I	W	O	F	P	U	X	O	G	H	C	E	N	M
G	Z	T	W	S	O	G	T	Q	A	G	P	T	E	Q
K	T	D	A	I	I	A	P	R	I	W	W	E	G	D
N	D	R	A	T	L	T	G	I	V	V	H	M	A	N
E	Z	H	L	U	S	E	I	G	R	U	N	B	T	H
P	X	R	S	Q	M	O	U	V	K	N	Z	M	I	Y
J	Q	N	V	K	L	I	R	H	E	O	N	O	V	Q
Z	I	Z	B	C	S	M	L	T	E	T	S	L	E	C
S	F	N	F	P	U	D	L	Z	C	U	C	U	E	J
E	R	O	F	D	Q	L	Q	U	Z	E	U	O	O	B
E	L	E	C	T	R	O	N	E	I	N	L	C	V	Z
D	L	E	I	F	N	E	U	A	V	A	O	E	B	Y
G	O	M	Q	J	X	O	I	T	N	T	D	I	M	E
K	Q	C	L	Q	X	P	F	T	S	B	K	P	Y	C

1. a. On the following three diagrams, what do the arrows show?

   e.g. an electron 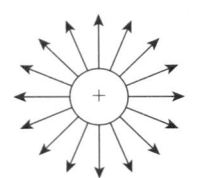   e.g. an ion 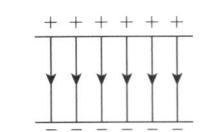   e.g. parallel charged plate

   .... [1]

   b. Draw the field lines between the two point charges. [3]

   + •   • −

80

Electric charge

11.3 Conductors and insulators

Language lab

Match the beginnings of the sentences with their endings.

Beginnings:

A positively charged object

A negatively charged object

Charge is transferred between objects by

The arrows on electric field lines show the

Endings:

has gained electrons.

direction a positive charge would move.

has lost electrons.

the force of friction. [4]

1. A student moves a charged polythene rod near to some very small pieces of paper. Even though the paper is not charged, the rod attracts the paper towards it. Draw a diagram showing the charges in the rod and the paper and explain why the paper is attracted to the rod.

   ....

   .... [3]

2. a. Aeroplanes move at great speed through the air. Explain how they become electrostatically charged.

   ....

   ....

   .... [3]

   b. When refuelling, an aeroplane is attached by a metal cable to the ground. Explain why.

   ....

   ....

   ....

   .... [3]

Electric charge

11.4 Charge and current

Language lab

Anagrams: unjumble the key words.

cur tern .... me pear ....

mere mat .... sir see .... [4]

1. a. i. Define electrical current.

   ....

   .... [1]

   ii. What is the unit of current? .... [1]

   iii. Which meter is used to measure current? Draw its symbol. [1]

   b. With reference to charges, describe current flow in a metal.

   ....

   .... [1]

   c. i. A current of 2.0 A flows in a circuit. How much charge passes a point in the circuit in 10 s?

   charge = ... [2]

   ii. Calculate the charge passing a point in a circuit in 3 s if the current flowing is 10 mA.

   charge = ... [2]

   iii. How much charge passes a point in a circuit in 30 minutes if the current flowing is 0.05 A?

   charge = ... [2]

   iv. How much current is flowing in a circuit if 120 C of charge pass a point in 1 minute?

   current = ... [3]

   v. What is the current flowing in a circuit if 18 mC pass a point in 6 s?

   current = ... [3]

   vi. How long does it take for 100 C to pass a point in a circuit when a current of 2 A flows?

   time = ... [2]

   vii. How long does it take for 10 mC to pass a point in a circuit when a current of 5 mA flows?

   time = ... [3]

Electric charge — Multiple choice questions

1. What is the effect of a negative charge on a positive charge?

   A   attract

   B   repel

   C   move at right angles upwards

   D   nothing

2. What are the charge carriers in a metal?

   A   protons

   B   neutrons

   C   electrons

   D   ions

3. What current flows if a charge of 0.06 C passes a point in 3 s?

   A   0.18 A

   B   0.02 A

   C   50 A

   D   1.8 A

4. What is the time taken for 0.5 C to pass a point if a current of 0.05 A flows?

   A   0.0025 s

   B   0.1 s

   C   1 s

   D   10 s

Electrical energy

12.1 Batteries and cells

Language lab

Complete the sentences using the key words.

complete   energy   negatively   current   electrical   chemical
voltage   positively   cells   potential difference

Batteries range in size from the tiny, low .... batteries found in calculators to the large batteries used to start a car. A battery consists of two or more .... connected together. Batteries transform .... energy to .... energy. One side of the battery is .... charged and the other side is .... charged. In a .... circuit, the opposite charges of the battery cause charges in the circuit to move and hence a .... flows. The electromotive force of a cell is measured in volts and is the maximum .... that a cell can supply. The higher the electromotive force of a cell, the greater the .... given to each charge as it passes through the cell. [10]

1. a. Describe an experiment to investigate how varying the number of cells connected in a series affects the potential difference across the cells. Include details of the independent, dependent, and controlled variables in the experiment, and the measuring instruments used.

   ....
   ....
   ....
   ....
   ....
   .... [5]

   b. Sketch a graph of what you expect your results to look like.

   [4]

Electrical energy   12.2 Potential difference

Language lab

Anagrams: unjumble the key words.

valet go ....

confidential free pet ....

revolt met ....

ratty be .... [4]

1. a. i. Define potential difference (voltage).

   .... [1]

   ii. What is the unit of potential difference? .... [1]

   iii. Which meter is used to measure potential difference? .... Draw its symbol below. [1]

   b. Three cells each of e.m.f. 1.5 V are joined in series as shown. What is the total e.m.f. that they can supply?

   e.m.f. = ... [1]

   c. In the space below, draw a circuit containing three identical bulbs in series and a battery of 6V. Indicate the potential difference across each bulb. [2]

Electrical energy — 12.3 Resistance

Language lab

Anagrams: unjumble the key words.

resort is ....   rec turn ....

insects ear ....   polar portion .... [4]

1. **a.** A student sets up the circuit shown (right).

   Describe how he can use the apparatus to investigate how current through a resistor varies with the potential difference across it.

   ....

   ....

   ....

   .... [4]

   b. The student plots a graph of his results. He has made three errors in presenting his data in this graph. What are the errors?

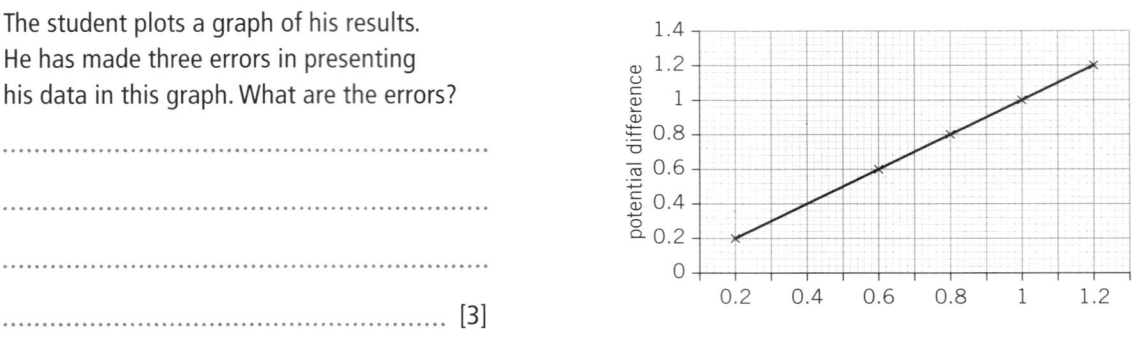

   ....

   ....

   ....

   .... [3]

   c. i. Use the graph to find the current through the resistor when the potential difference (p.d.) is 0.45 V.

   current = ... [1]

   ii. Use the graph to find the current when the p.d. is 0.9 V.

   current = ... [1]

   d. The student suggests that current is directly proportional to voltage. Do the answers to **c.** and **d.** support this suggestion? Explain your answer.

   ....

   .... [2]

   e. A second student comments that this relationship between current and voltage is only true for the range of voltages tested and that the graph would no longer be a straight line at higher voltages.

   Explain whether the second student is correct.

   ....

   .... [2]

   f. By referring to its atomic structure, describe how a resistor provides resistance to current flow.

   ....

   ....

   .... [3]

Electrical energy

12.4 More about resistance

Language lab

Match the key words with their meanings.

Key word	Meaning
resistance	used to measure current in a circuit
current	potential difference required to make one ampere of current flow
potential difference	the charge passing a point per second in a circuit
ammeter	a source of chemical energy that causes current to flow around a circuit
voltmeter	the work done on a coulomb of charge between two points in a circuit
cell	used to measure potential difference in a circuit

[6]

1. A student predicts that the resistance of a piece of wire increases with the length of the wire. She sets up a circuit with a piece of constantan wire, a battery, a voltmeter, and an ammeter.

   a. Draw a circuit diagram of the apparatus that could be used to test the student's prediction.

   [3]

   b. Suggest a measuring instrument she could use to measure length. .... [1]

   The student obtains the following results from her experiment:

length / cm	voltage / V	current / A	resistance /
10.0	2.0	2.00	
20.0	2.0	0.98	
30.0	2.0	0.68	
40.0	2.0	0.50	
50.0	2.0	0.41	
60.0	2.0	0.33	
70.0	2.0	0.28	

   c. Complete the final column in the table with the unit and values of resistance. [3]

   d. Was the student's prediction correct? Explain your answer with evidence from the table.

   .... [3]

2. a. What is the equation that relates resistance, current, and potential difference (voltage)?

   .... [1]

   b. Draw the circuit symbol for a resistor. [1]

   c. i. What is the potential difference across a 6 Ω resistor when a current of 3 A flows through the resistor?

   potential difference = ... [2]

   ii. What is the value of a resistor if a potential difference of 12 V causes a current of 0.5 A to flow through it?

   resistance = ... [2]

87

Electrical energy — 12.5 Electrical power

Language lab

Complete the sentences using the key words, which can be used more than once.

    ions   power   vibrate   energy   circuit   light

Work is done on charges as they pass through a cell and gain .... As the charges pass through a bulb in the .... , they pass their .... to the metal .... in the bulb. The ions .... more and the filament becomes hotter. The bulb gives out .... and heat energy. The work done per second on the bulb is the ....

[7]

1. a. i. Write down an equation that relates electrical power, current, and potential difference. [1]

   ii. What is the unit of power?... [1]

   b. i. A bulb of power 40 W has a potential difference of 12 V across it. What is the current flowing in the bulb?

   current = ... [2]

   ii. What is the power of a bulb that allows a current of 2 A to flow when there is a potential difference of 10 V across it?

   power = ... [2]

   iii. What potential difference is required to produce a current of 0.5 A in a bulb of power 60 W?

   potential difference = ... [2]

Electrical energy — Multiple choice questions

1. What is the resistance of a bulb if a current of 0.5 A flows through it when there is a p.d. of 12 V across it?

    A 2.4 Ω
    B 6 Ω
    C 0.6 Ω
    D 24 Ω

2. What is the p.d. across a bulb of resistance 100 Ω when a current of 0.02 A flows?

    A 2 V
    B 5000 V
    C 500 V
    D 0.0002 V

3. What is the power of a bulb if a current of 5 A flows when there is a p.d. of 12 V across it?

    A 60 W
    B 2.4 W
    C 30 W
    D 6 W

4. What is the p.d. across a bulb of power 30 W when a current of 0.4 A flows through it?

    A 1.5 V
    B 230 V
    C 75 V
    D 12 V

Electric circuits

13.1 Circuit components

Language lab

Anagrams: unjumble the key words.

hermit sort ....

a laborer revisits ....

deed interpret slingshot ....

die do .... [4]

1. a. Complete the table by naming the components. [6]

symbol	component	symbol	component
(thermistor symbol)		(capacitor symbol)	
(LED symbol)		(variable resistor symbol)	
(LDR symbol)		(resistor symbol)	

b. A student carries out an experiment to find out how the resistance of a thermistor varies with temperature. The student has the following equipment:

A thermistor, thermometer, battery, beaker, kettle, ice, ammeter, voltmeter, battery, leads to connect the circuit.

Draw a labelled diagram and a circuit diagram to show how the student should set up the apparatus.

[4]

c. Describe how the student should use the apparatus to find out how the resistance of the thermistor varies with temperature.

....

....

....

.... [4]

Electric circuits — 13.2 Series circuits

Language lab

Complete the sentences using the key words.

    ammeter   **voltmeter**   **current**   **parallel**   **variable resistor**

In a circuit, a .... is placed in .... with the component to measure the potential difference, and an .... is placed in series with the component, to measure the current. The potential difference across the component can be varied using a .... It is often necessary to limit the amount of .... in a circuit to avoid the risk of fires. [5]

1. A student carries out an experiment to vary the number of bulbs in series in a circuit and measure the current in the circuit.

   a. Draw a circuit diagram for the experimental arrangement. [3]

   b. Suggest one variable the student should keep controlled in the experiment.

   .... [1]

   The student obtains the following results:

number of bulbs	1	2	3	4	5	6
current / A	2.00	0.98	0.65	0.51	0.40	0.32

   c. The student states that the relationship between the two variables is as follows:

   'The current in the circuit is inversely proportional to the number of bulbs.'

   Use the data in the table to provide evidence to support or reject the student's conclusion.

   ....

   .... [3]

   d. Explain why there is this relationship between current and number of bulbs.

   ....

   .... [3]

Electric circuits

13.3 Parallel circuits

Language lab

Match the key words with their definitions.

independent variable	the variable that is measured in an experiment
dependent variable	the variable that is kept the same in an experiment
controlled variable	the variable that is changed in an experiment [3]

1. A student wishes to investigate how the number of resistors in parallel affects the current in a circuit. She has a number of 10 Ω resistors, a power supply, and an ammeter.

   a. Draw a circuit diagram for the experiment. [3]

   b. Suggest one variable that must be kept controlled during the experiment. .... [1]

   The student obtains the following results:

number of resistors	current / A	current / A	average current / A
1	0.20	0.20	
2	0.40	0.41	
3	0.58	0.61	
4	0.81	0.79	
5	1.00	0.99	

   c. Complete the table by calculating the average values of current. [2]

   d. Plot a graph of average current against number of resistors. [4]

   e. What is the relationship between the average current and the number of resistors in parallel?

   ....

   ....

   .... [2]

   f. Explain the relationship between the average current and number of resistors in parallel.

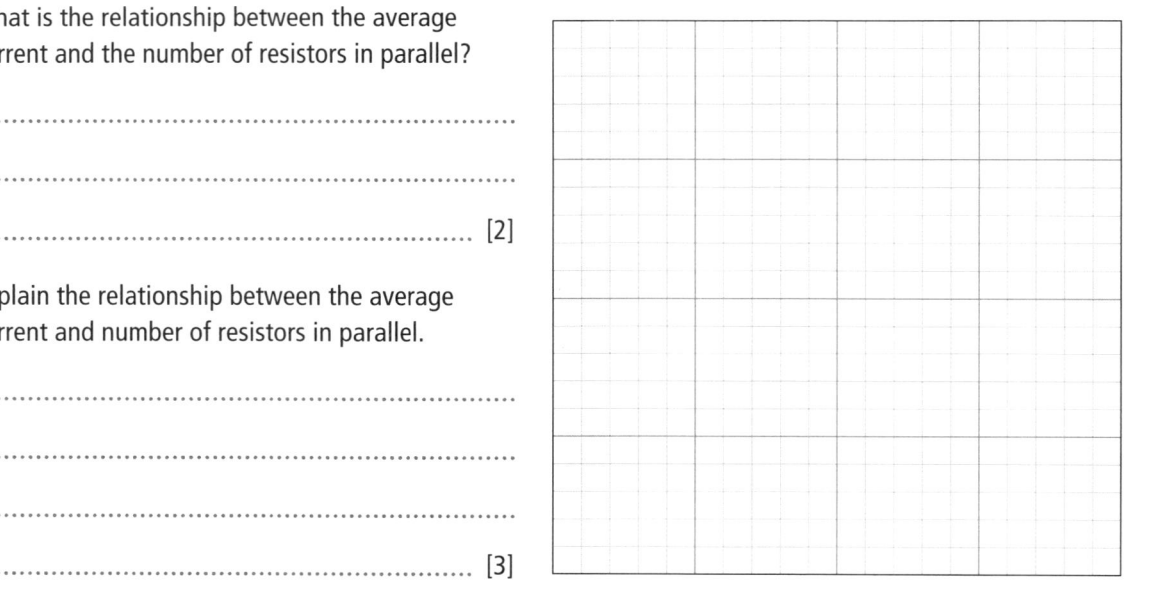

   ....

   ....

   .... [3]

92

Electric circuits

13.4 More about series and parallel circuits

Language lab

Anagrams: unjumble the key words.

see sir ....

real pall ....

sorriest ....

collective fleet moo .... [4]

1. a. For the following resistor networks, calculate the combined resistance. [3]

   - 5 Ω, 6 Ω
   - 120 Ω, 340 Ω
   - 5 Ω, 10 Ω, 10 Ω

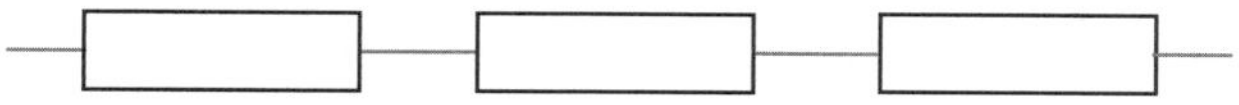

   b. Find the missing values of current and potential difference in the following circuits. [8]

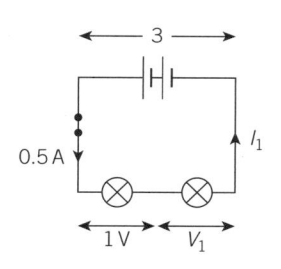

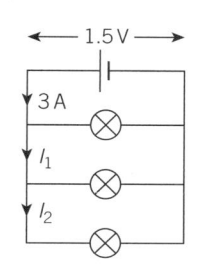

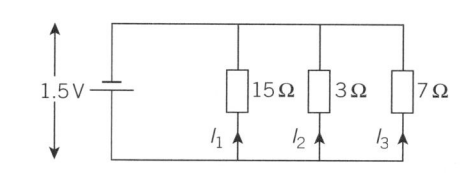

   Circuit 1: 3 V battery, 0.5 A, 1 V, V_1, I_1

   I_1 = ...

   V_1 = ...

   Circuit 2: 1.5 V, 3 A, I_1, I_2, V_1

   I_1 = ...

   I_2 = ...

   V_1 = ...

   Circuit 3: 1.5 V, 15 Ω, 3 Ω, 7 Ω, I_1, I_2, I_3

   I_1 = ...

   I_2 = ...

   I_3 = ...

   c. The following resistor network is connected across a battery. State and explain which of the three resistors carries the largest current.

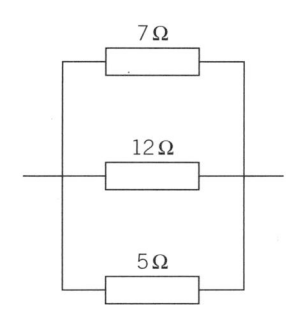

   7 Ω, 12 Ω, 5 Ω

   ....

   ....

   ....

   .... [3]

93

Electric circuits

13.5 Sensor circuits

Language lab

Complete the crossword.

Across

2. Stores chemical energy and transforms it to electrical energy
3. Adding cells in series creates a ....
6. Its resistance decreases as the light intensity increases
8. Uses a small potential difference to switch on a large potential difference
10. Only allows current to flow on one direction around a circuit
11. In a .... circuit, the current is the same at all points in the circuit

Down

1. What e.m.f. stands for
3. An electrical component that emits light
4. Its resistance can be changed in order to vary the potential difference across a component
5. In a .... circuit, the current splits at the junctions
7. Its resistance decreases as temperature increases
9. Breaks a circuit when the current becomes too large

[11]

1. From the following list, sort the components into input sensors and output devices.

| microphone | LDR | loudspeaker | relay | pressure switch | buzzer | heater |
| variable resistor | LED | lamp | thermistor | reed switch | electric bell | |

input sensors	output devices

[4]

2. Explain the difference between an analogue and a digital signal.

....

.... [2]

Electric circuits

13.6 Switching circuits

Language lab

Match the key words with their meanings.

thermistor	transforms chemical to electrical energy
fuse	uses a smaller current to switch on a larger current
variable resistor	has a resistance that decreases as temperature increases
cell	melts and breaks the circuit if the current is too high
LED	an electrical switch operated by a magnetic field
Reed switch	has a resistance that decreases as light intensity increases
relay	used to vary the resistance in a circuit
LDR	a diode that emits light [8]

1. The circuit diagram below shows a potential divider circuit. The picture to the right shows the variable resistor in more detail.

   a. Name the unlabelled component in the circuit. .... [1]

   b. Describe how the value on the voltmeter varies as the slider on the variable resistor is moved from point 1 to point 2.

   ....

   .... [2]

2. A student sets up the circuit shown (right).

   a. Label the components in the circuit. [3]

   b. The student is surprised to find that the bulb does not light.
      State and explain how the circuit should be changed so that the bulb will light.

   ....

   ....

   .... [2]

   c. The student builds a second circuit, shown right. Which, if any, of the bulbs **A**, **B**, **C** will light? [1]

      Explain your answer.

      ....

      .... [2]

95

Electric circuits

13.7 Logic circuits

Language lab

Complete the sentences using the key words.

| NOT | logic | output | low | digital | high |

Logic gates are used to build .... circuits. .... gates usually have two inputs and one .... but the .... gate has one input and one output. The inputs and output have two possible states .... (0) or .... (1), which represent different voltage levels. [6]

1. Name the logic gates shown and write their truth tables. [10]

a. 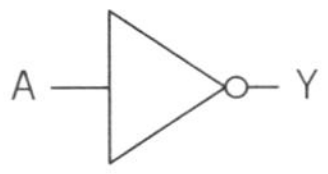 A —▷o— Y

b. 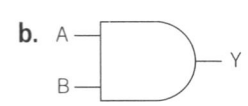 A, B —D— Y

c. 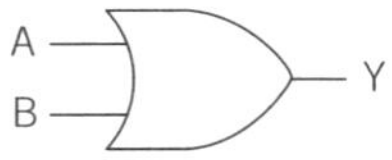 A, B —)— Y

d. A, B —D o— Y

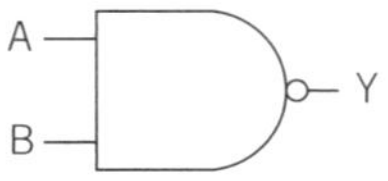

e. 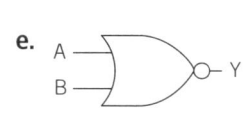 A, B —)o— Y

96

Electric circuits

13.8 Logic circuits in control

Language lab

Match the beginnings of the sentences to the endings.

Beginnings:

An AND gate gives a high output

An OR gate gives a low output

A NOT gate gives a high output

An OR gate gives a high output

Endings:

if the input is low.

if both the inputs are high.

if both the inputs are low.

if one or both inputs are high. [4]

1. a. The diagram below shows a digital circuit made from three NOT gates and one NAND gate.

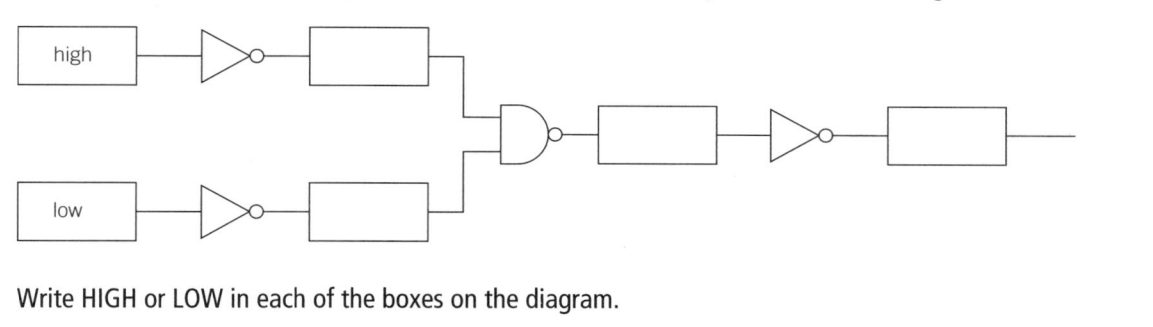

   Write HIGH or LOW in each of the boxes on the diagram. [4]

   b. State the effect on the output of changing the high input to low. .... [1]

Electric circuits

13.9 Electrical safety

Language lab

Anagrams: unjumble the key words.

cacti saplings ....

sufe ....

ritual renew ....

rehear wit .... [4]

1. The diagrams below show a three-pin and a two-pin plug with some of the safety features labelled.

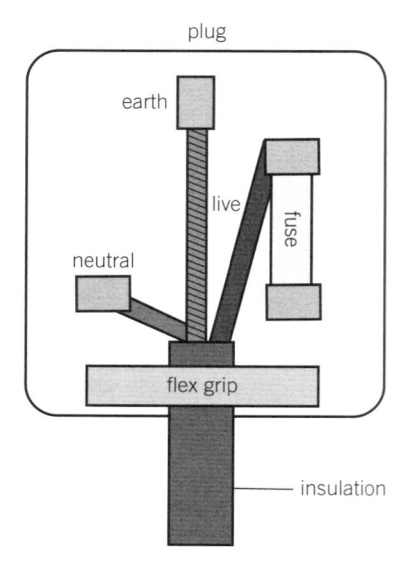

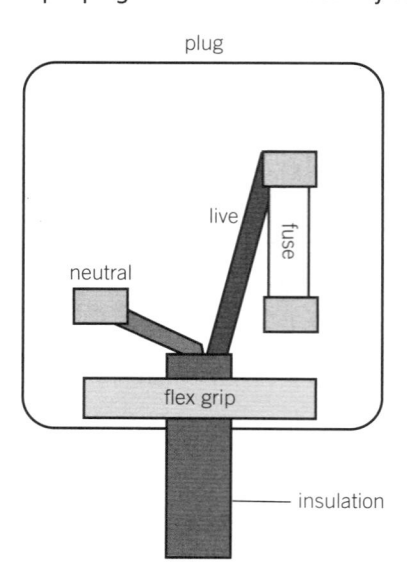

   a. State the importance of:

      i. the flex grip

      ....

      .... [2]

      ii. the fuse

      ....

      .... [2]

      iii. the plastic plug cover.

      ....

      .... [1]

   b. Explain the action of the earth wire in protecting people from electric shocks.

      ....

      ....

      .... [3]

Electric circuits

13.10 More about electrical safety

Language lab

Complete the sentences using the key words. You will not need to use all the words provided.

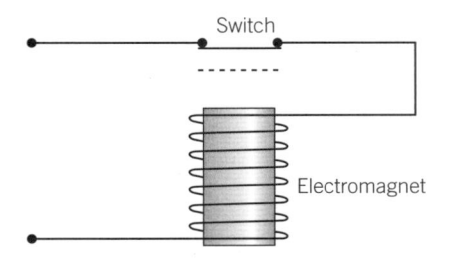

Hz   fuse   zero   large   fifty   neutral   melting   potential difference   live   positive   earth

A mains circuit consists of a live wire and a .... wire. The .... between live and earth alternates between .... and negative. In most countries around the world, the voltage alternates .... times per second (at a frequency of 50 ....). The potential difference between neutral and .... is close to .... A .... in the live wire protects against fires by melting and breaking the circuit if the current becomes too .... [9]

1. a. The photograph and diagram above show a circuit breaker found in household circuits. Explain how a circuit breaker protects against house fires.

   ....
   ....
   ....
   .... [3]

b. An electrical device with a plastic case is said to have 'double insulation'. Explain what this means.

   ....
   ....
   .... [2]

Electric circuits — Multiple choice questions

1. Which component can be used to vary the potential difference across another component in a circuit?

   A   resistor
   B   variable resistor
   C   bulb
   D   thermistor

2. Which component has a resistance which varies with incident light intensity?

   A   thermistor
   B   resistor
   C   LDR
   D   bulb

3. Which component only allows current to flow through it in one direction?

   A   diode
   B   variable resistor
   C   fuse
   D   thermistor

4. Which component can be used to switch on a large p.d. using a small p.d.?

   A   thermistor
   B   resistor
   C   cell
   D   relay

5. Which of the following statements about resistors in circuits is **correct**?

   A   In a series circuit the current is shared between resistors.
   B   In a series circuit the potential difference is shared between resistors.
   C   In a parallel circuit the potential difference is shared between resistors.
   D   In a parallel circuit the current through every resistor is the same.

6. Which feature of a plug protects the user from fires?

   A   fuse
   B   earth wire
   C   plastic case
   D   cable grip

7. Which feature of a 3-pin plug is missing from a 2-pin plug?

   A   fuse
   B   earth wire
   C   plastic case
   D   cable grip

8. Which of the following devices can be used as an input sensor in an electronic circuit?

   A   loudspeaker
   B   relay
   C   microphone
   D   lamp

Electric circuits — Multiple choice questions

9. Which of the following devices can be used as an output device in an electronic circuit?

   A lamp   B LDR

   C thermistor   D switch

10. The following truth table is for which logic gate?

Input	Input	Output
0	0	0
0	1	0
1	0	0
1	1	1

   A NOT   B AND

   C NOR   D NAND

11. The following truth table is for which logic gate?

Input	Input	Output
0	0	1
0	1	0
1	0	0
1	1	0

   A NOT   B AND

   C NOR   D NAND

12. The following is the symbol for which logic gate?

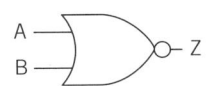

   A NOT   B AND

   C NOR   D NAND

Electromagnetism

14.1 Magnetic field patterns

Language lab

Anagrams: unjumble the key words.

airman router .... layer ....

nail tours .... rectangle tome ....

[4]

1. Draw the magnetic field lines formed when a current flows in a wire, as shown in the diagram. [2]

electric current →

2. Explain how the relay shown below can use a small potential difference in one circuit to switch on a high potential difference in another circuit.

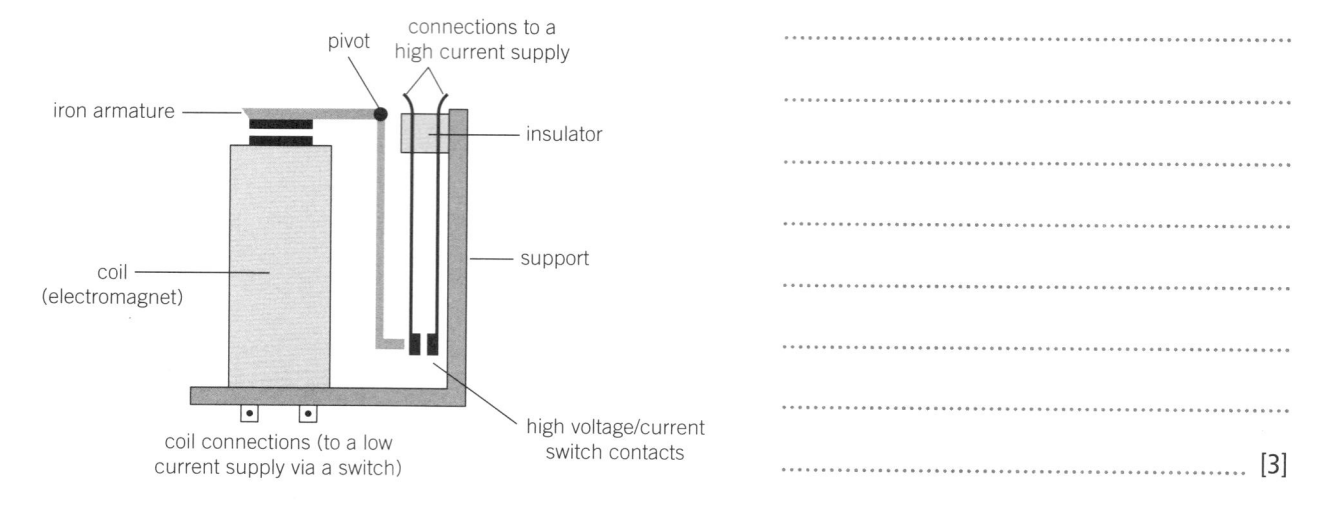

....

....

....

....

....

....

.... [3]

Electromagnetism

14.2 The motor effect

Language lab

Complete the sentences using the key words.

| magnetic field | size | current | direction | strength | force |

When a wire carrying a current is placed in a magnetic field, there is a .... on the wire. The size of the force is dependent on the .... of the .... and the .... of the .... The .... of the force is dependent on the direction of the current and the direction of the magnetic field. [6]

1. A wire carrying a current is placed between the poles of a magnet and the wire experiences a force, as shown in the diagram.

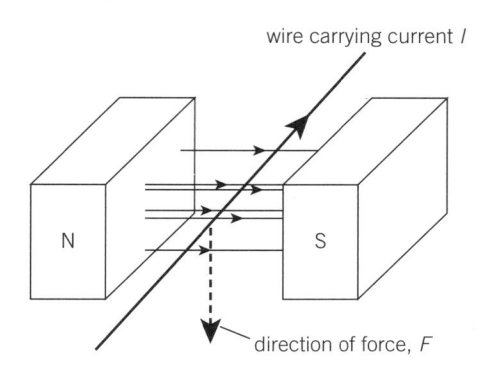

   a. Describe the rule that predicts which way the wire will move in the magnetic field.

   ....
   ....
   .... [3]

   b. What will happen to the original wire if:

   i. The magnetic poles are reversed.

   .... [1]

   ii. The current flows in the opposite direction.

   .... [1]

   iii. The current is decreased.

   .... [1]

   iv. The strength of the magnetic field is increased by using a stronger magnet.

   .... [1]

Electromagnetism   14.3 The electric motor

Language lab

Complete the word search.

Words may go downwards, upwards, forwards, backwards, or diagonally.

- brush
- coil
- commutator
- current
- force
- magnet
- motor
- pole

M	V	V	D	B	S	V	F	L	R	F	R	B	F	A
F	J	V	B	A	O	U	J	B	O	O	D	Y	M	J
T	A	T	C	Z	U	E	F	R	T	J	T	N	U	X
B	L	H	U	I	T	C	C	A	D	X	D	O	M	Y
T	Z	T	Y	Z	H	E	T	F	V	R	C	C	M	Q
A	W	O	G	J	M	U	Z	L	Z	H	S	U	R	B
M	K	E	J	W	M	A	W	T	N	Z	G	Y	I	O
Q	A	P	Y	M	E	U	K	X	K	C	A	T	A	D
L	B	G	O	N	C	U	R	R	E	N	T	P	M	A
Q	H	C	N	L	M	D	I	M	K	B	U	O	A	Y
A	R	P	I	E	T	H	Z	K	C	D	Z	Y	C	M
Z	C	O	R	B	T	X	S	V	X	P	H	E	O	L
V	C	C	S	N	E	J	D	W	S	U	G	L	N	Y
K	Q	C	L	L	A	P	F	S	P	W	P	O	D	G
M	F	I	O	F	O	W	J	N	I	K	J	P	W	K

[8]

1. The diagram below shows a d.c. motor.

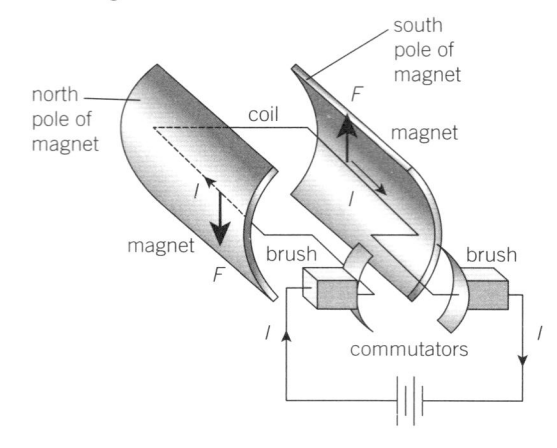

   a. What would be the effect on the speed of rotation of the coil if:

   i. The strength of the magnetic field was increased? .... [1]

   ii. The current was increased? .... [1]

   iii. A cell was removed from the battery? .... [1]

   iv. The number of turns on the coil was increased? .... [1]

   b. Explain the function of the pole magnet pieces and why they are curved.

   ....

   .... [2]

   c. Explain the function of the brushes and suggest a material from which the brushes could be made.

   ....

   ....

   .... [3]

Electromagnetism — 14.4 Electromagnetic induction

Language lab

Use the key words to complete the sentences.

alternating   magnet   complete   coil   direction   potential   difference

When a .... is moved into and out of a coil, a potential difference is induced across the .... The direction of the .... changes as the .... of motion of the magnet changes. If there is a .... circuit, an .... current will flow. [7]

1. A student sets up the apparatus shown below. She moves the magnet into the coil and records the observations shown below.

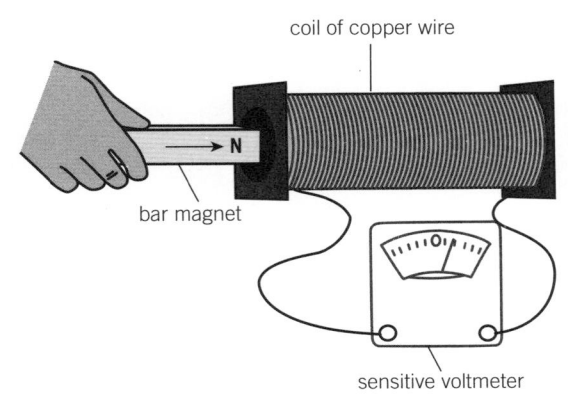

action	observation
move north pole slowly into the coil	small positive deflection on the voltmeter
move north pole quickly into the coil	large positive deflection on the voltmeter
move north pole quickly out of the coil	large negative deflection on the voltmeter
move south pole slowly into the coil	
move south pole quickly into the coil	
move south pole quickly out of the coil	

a. Complete the table of observations. [3]

b. State the effect on the observations in the table if the coil is replaced with one of double the number of turns.
.... [1]

c. Explain how the deflection on the voltmeter is produced.
....
....
.... [3]

d. Explain why both negative and positive deflections are produced.
....
.... [2]

105

Electromagnetism

14.5 The alternating current generator

Language lab

Match the beginnings of the sentences with their endings.

Beginnings:

When the coil spins it

When the coil cuts the field lines, a potential

As there is a complete circuit,

The sides of the coil alternately cut down and

Endings:

difference is induced across the coil.

then up through the field lines.

cuts the magnetic field lines due to the magnet.

an alternating current flows in the coil. [4]

1. The diagram below shows an a.c. generator.

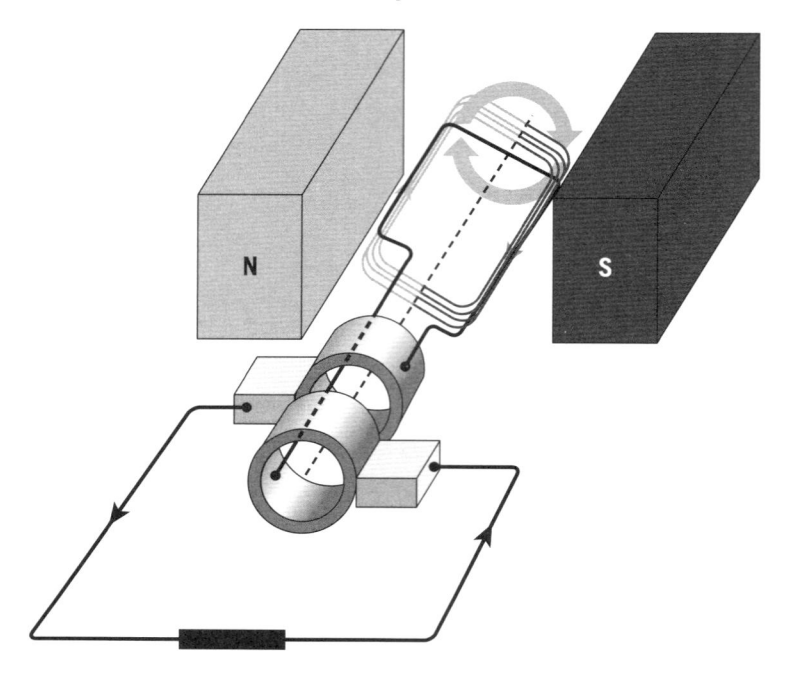

    a. State the difference between alternating and direct current.

    ....

    ....

    .... [2]

    b. The coil is spun as shown. On the diagram, add a voltmeter in parallel with the resistor. [1]

    c. Explain the purpose of the brushes and slip rings.

    ....

    ....

    .... [2]

    d. Describe how the needle on the voltmeter will move as the coil spins.

    ....

    ....

    .... [2]

Electromagnetism   14.6 Transformers

Language lab

Anagrams: unjumble the key words.

reforms rant .... decoys ran ....

prim ray .... integral ant ....

[4]

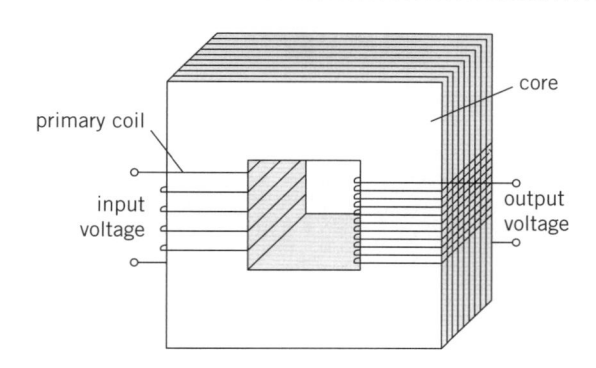

1. a. Write down the equation that links the voltage (potential difference) in the primary and secondary coils of a transformer and the number of turns on the coils.

   .... [1]

   b. Calculate the output voltage when 12 V is applied to a primary coil of 50 turns and the secondary coil has 100 turns.

   output voltage = ... [3]

2. a. Put the following statements in order to explain how the transformer shown below steps up potential difference from a low value to a high value. The first one has been done for you.

statement	order
An alternating voltage is applied to the primary coil.	
The changing magnetic field in the core induces a voltage in the secondary coil.	
The alternating current in the primary coil creates an alternating magnetic field around the coil.	
This causes an alternating current to flow in the primary coil.	
The voltage across the secondary coil (the output voltage) is greater than that across the primary because there are more turns on the secondary coil.	
The core becomes magnetised and its magnetic field also alternates.	

[6]

   b. Explain how the step-up transformer shown could be adapted to make a step-down transformer.

   ....
   .... [2]

   c. A step-down transformer has an output voltage of 12 V. There are 50 turns on the secondary coil and 1000 turns on the primary coil. Calculate the primary voltage.

   primary voltage = ... [3]

107

Electromagnetism — 14.7 High-voltage transmission of electricity

Language lab

Complete the crossword.

Across

3. The material from which a transformer core is made
6. In a transformer, the output potential difference is induced across this coil
8. When a potential difference is produced across a conductor due to a changing magnetic field
11. In a step-up transformer, there are .... turns on the secondary coil than on the primary
12. Acts on a wire carrying a current in a magnetic field

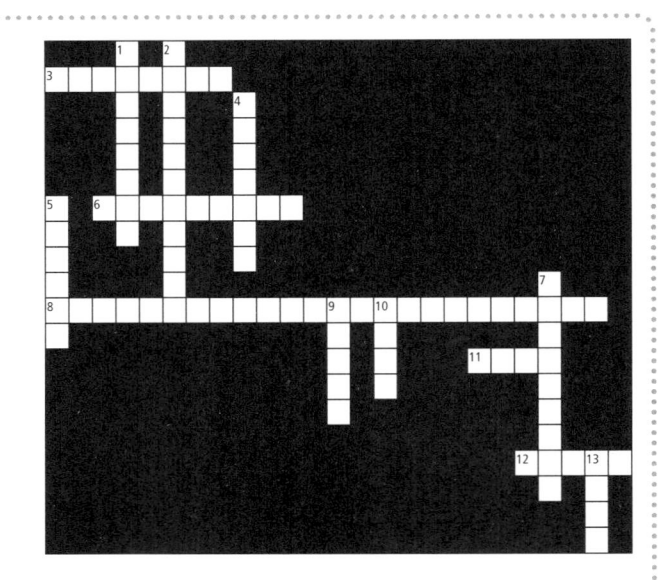

Down

1. Using a .... magnet can increase the induced potential difference
2. Used to step up and step down potential difference
4. In a transformer, the input potential difference is across this coil
5. Moving the magnet .... can increase the induced potential difference
7. Changing the .... of motion of a magnet into a coil changes the direction of the induced potential difference
9. Having a coil with more .... can increase the induced potential difference
10. When a coil .... magnetic field lines, a potential difference is induced
13. This object spins due to the magnetic field inside a d.c. motor

[13]

1. a. With reference to energy, explain why transformers are used to step up the potential difference from power stations onto the power lines.

   ....
   ....
   ....
   .... [4]

   b. Explain why mains electricity is produced and distributed as alternating current.

   ....
   ....
   .... [3]

   c. State one advantage and one disadvantage of underground power lines compared to those suspended on pylons above ground.

   ....
   .... [2]

Electromagnetism — Multiple choice questions

1. Which rule predicts which way a current-carrying wire inside a magnetic field will move?

    A   Fleming's right-hand rule

    B   Left-hand grip rule

    C   Fleming's left-hand rule

    D   Right-hand rule

2. Which of the following is **not** a method of increasing the induced e.m.f. in a coil when a magnet is moved into and out of the coil?

    A   Increase the number of turns on the coil

    B   Increase the strength of the magnet

    C   Move the magnet more quickly

    D   Turn the magnet round

3. What is the voltage across the secondary coil of a transformer if the primary voltage is 12 V, there are 100 turns on the primary, and 300 turns on the secondary?

    A   4 V

    B   36 V

    C   1200 V

    D   6 V

4. Which of the following is **not** a way to increase the speed at which a d.c. motor turns?

    A   Increase the number of turns on the coil

    B   Increase the magnetic field

    C   Increase the potential difference

    D   Decrease the current

Radioactivity

15.1 Observing nuclear radiation

Language lab

Match the key words with their meanings.

protactinium	an isotope with unstable nuclei that emit radioactive particles
radioisotope	the contribution of building materials, food etc. to the detected radiation
half-life	a radioisotope
background count	the time taken for half of the radioactive nuclei to decay [4]

1. A teacher uses a Geiger–Muller tube and counter to detect radiation from a radioactive source, as shown in the diagram (right).

   She wishes to demonstrate to her IGCSE students how to determine which type or types of radiation are emitted by the source.

   a. Describe three safety precautions the teacher must undertake to keep her and her students safe during the experiment.

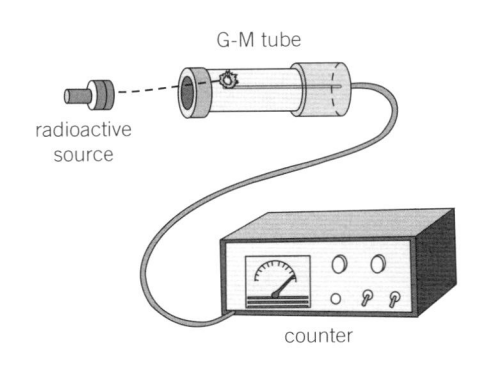

   ....

   ....

   ....

   .... [3]

   b. Before beginning the experiment she takes five readings of background radiation, each over a period of 30 seconds, and records the following counts:

   $$15, 16, 13, 10, 18$$

   Calculate the average count over 30 seconds.

   average count = ... [1]

   c. The teacher places a piece of paper, a thin sheet of aluminium, and a sheet of lead in turn between the detector and radioactive source. The students record the following results:

absorber	count over 30 s without the absorber	(count without the absorber) − minus (average background count)	count over 30 s with the absorber	(count with the absorber) − minus (average background count)	difference in corrected count rate over 30 s
paper	333		326		
aluminium	341		164		
lead	321		67		

   Complete the table by subtracting the average background count (which you calculated in part **b.**) from the count over 30 s. For each material, calculate the difference in count, corrected for background radiation, over 30 seconds with and without each of the absorbers.

   d. Which types of radiation are emitted by the source .... [1]

Radioactivity

15.2 Alpha, beta, and gamma radiation

Language lab

Match the key words with their definitions.

radioactive	changed direction
alpha	can get through many materials
beta	collides with and removes electrons from atoms
gamma	a helium nucleus
ionising	a fast-moving electron
penetrating	an electromagnetic wave of high energy
deflected	has an unstable nucleus and emits radioactive particles [7]

1. a. Name the three types of radiation that can be emitted from a radioactive nucleus.

   .... [3]

   b. Complete the table about the three types of radiation.

type of radiation			
what is it?	helium nucleus	fast-moving electron	
relative charge		−1	
relative mass			0

   [8]

   c. Why does a radioactive nucleus emit radiation?

   .... [1]

   d. All three types of radiation are ionising. What is the effect of ionising radiation on atoms?

   .... [1]

   e. i. Put the three types of radiation in order from least ionising to most ionising.

   least ionising ⟶ most ionising   [1]

   ii. Put the three types of radiation in order from least penetrating (least able to pass through materials) to most penetrating.

   least penetrating ⟶ most penetrating   [1]

   f. Radiation can be deflected by an electric field. The diagram shows the paths of the three types of radiation in an electric field. Label the paths with each type of radiation. [3]

111

Radioactivity

15.3 The discovery of the nucleus

Language lab

Anagrams: unjumble the key words.

ethical rap pal .... ceded left ....

golf idol .... aviator dice .... [4]

1. The diagram below shows Rutherford's alpha scattering experiment. A beam of positively charged alpha particles from a radioactive source were fired at a very thin gold foil and the deflected particles detected after they had passed through the foil.

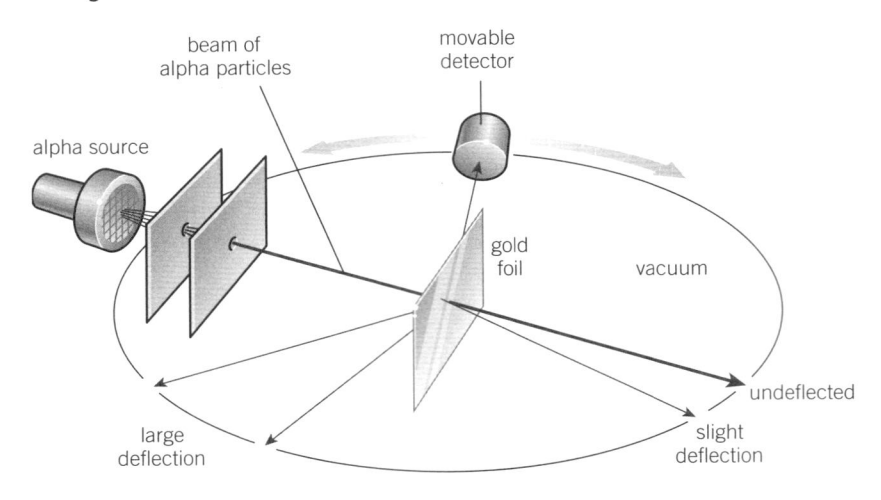

   a. i. Why was the apparatus placed in an evacuated vessel?

   .... [2]

   ii. Why was the detector movable?

   .... [1]

   iii. Why was a very fine beam of alpha particles used?

   .... [2]

   b. i. Most of the alpha particles passed straight through the gold foil undeflected. What did this tell Rutherford about the structure of the atom?

   .... [1]

   ii. Some alpha particles were deflected through small angles. Explain what caused the force on the alpha particles that made them change direction.

   ....

   .... [2]

   iii. Very few alpha particles were deflected through large angles, up to 180°. Explain how this led to the conclusion that the nucleus is very small compared to the size of the atom.

   ....

   .... [2]

112

Radioactivity

15.4 More about the nucleus

Language lab

Match the key words with their meanings.

atom	the path taken by an electron around the nucleus
electron	an uncharged particle found in the nucleus
proton	an atom containing the same number of protons but a different number of neutrons
neutron	the central part of the atom
nucleus	a negatively charged particle of very small mass
orbit	a tiny particle containing a nucleus and electrons
isotope	a positively charged particle found in the nucleus

[7]

1. The diagram represents an atom.

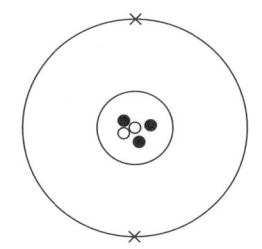

    a. Label the following on the diagram:
       i. the nucleus
       ii. the electrons
       iii. the protons
       iv. the neutrons. [4]

    b. Complete the table of relative charges and masses of the particles in an atom.

name of particle	relative mass	relative charge
proton		
neutron		
electron		

    [6]

    c. What is the general name for protons and neutrons? .... [1]

    d. What is an isotope?

    ....

    .... [2]

    e. Draw a diagram of an isotope of the element shown in **a.** above.

2. The nucleus of carbon-14 decays to nitrogen-14 by the emission of a particle.

    a. i. Complete the decay equation for carbon-14. [2]

    $$^{14}_{6}C \longrightarrow {^{14}_{7}}N + {^{-}_{-}}X$$

    ii. What is X? .... [1]

    b. i. Complete the decay equation for cobalt-60. [4]

    $$^{60}_{27}Co \longrightarrow {^{-}_{-}}X + \gamma$$

    ii. What is X? .... [1]

113

Radioactivity

15.5 Half-life

Language lab

Complete the sentences using the key words.

average	radiation	random	Geiger–Muller tube	accurate	subtracted
	counter		background count	source	

A .... and a .... are used to detect radiation from a .... Different types of .... are absorbed by different materials. The .... must be taken first so that this can be .... from the count from the source. Radioactive decay is a .... process and so it is important to take several readings and .... them to obtain a more .... value. [9]

1. A group of students monitor the count rate from a sample of the radioisotope protactinium.

    a. The students use graph plotting software to obtain the following graph of their data as protoactinium decays.

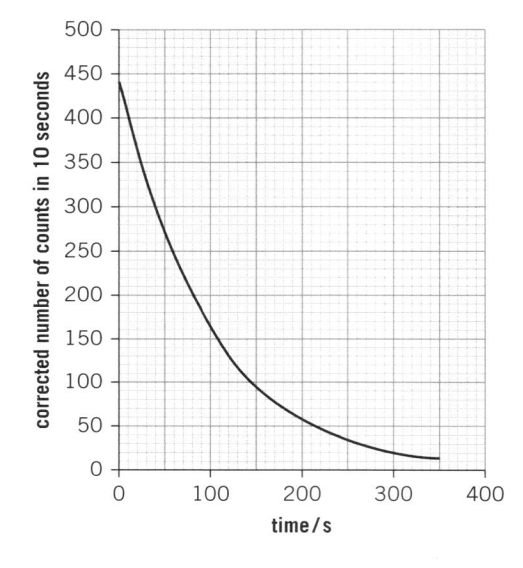

    i. What is meant by the half-life of a radioisotope?
    .... [1]

    ii. Use the graph to find the half-life of protactinium. Show your working here.

    half-life = ... [1]

    b. A sample of protactinium contains 100 million protactinium nuclei.

    i. How long will it take for the number of protactinium nuclei to be reduced to 25 million?

    time = ...

    ii. How long will it take for the number to be reduced to 6.25 million?

    time = ...

Radioactivity

15.6 Radioactivity at work

Language lab

Anagrams: unjumble the key words.

numb roper not	hall fife ....
semi ions	able stun ....
column been run	clue sun .... [6]

1. In a paper mill, a beta radiation source with a half-life of over 10 years is used to monitor the thickness of the paper.

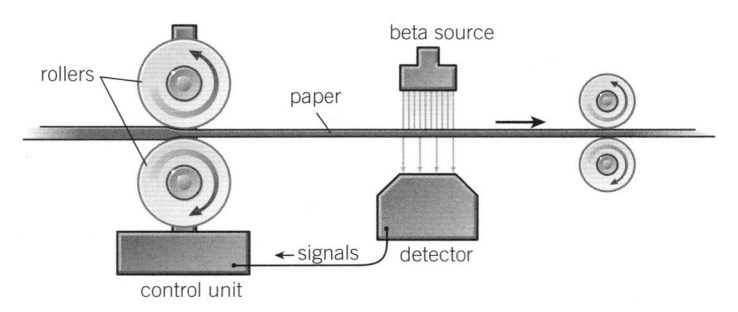

   a. Why is a beta source used instead of an alpha or gamma source?

   ....
   ....
   .... [3]

   b. Explain what happens to the count rate at the detector if the paper becomes too thick.

   ....
   .... [2]

   c. What does the control unit do if the paper becomes too thick?

   ....
   .... [1]

3. Leaks from water pipes buried underground can be detected by introducing a radioisotope into the water and then using a detector to measure the count rate above ground.

   a. Explain whether the radioisotope used should be an alpha, beta, or gamma emitter.

   ....
   .... [3]

Radioactivity — Multiple choice questions

1. Which type of radiation is a helium nucleus?

    A   alpha
    B   beta
    C   gamma
    D   X-ray

2. Which type of radiation is unable to penetrate 10 cm of air?

    A   alpha
    B   beta
    C   gamma
    D   X-ray

3. Which type of radiation is attracted to a positively charged metal plate?

    A   alpha
    B   beta
    C   gamma
    D   X-ray

4. Which type of radiation is attracted to a negatively charged metal plate?

    A   alpha
    B   beta
    C   gamma
    D   X-ray

5. Which two types of radiation are deflected in opposite directions in a magnetic field?

    A   alpha, gamma
    B   beta, gamma
    C   alpha, beta
    D   X-ray, gamma

6. Which are the positive charges in an atom?

    A   electrons
    B   protons
    C   neutrons
    D   ions

7. The nucleus of carbon-14 contains six protons. How many electrons are there in a neutral carbon atom?

    A   5
    B   6
    C   8
    D   14

8. Magnesium-25 contains 13 neutrons. How many electrons are there in a neutral atom?

    A   13
    B   12
    C   25
    D   11

9. A radioisotope has a half-life of 2 days. What proportion of the original isotope remains after 4 days?

    A $\frac{1}{2}$
    B $\frac{1}{4}$
    C $\frac{1}{8}$
    D $\frac{1}{16}$

Radioactivity — Multiple choice questions

10. A radioisotope decays to $\frac{1}{8}$ of its original activity in 6 hours. What is its half-life?

    A   6 h
    B   8 h
    C   1.5 h
    D   2 h

11. How long does it take for a radioisotope of half-life 3.2 s to decay to $\frac{1}{16}$ th of its original activity?

    A   3.2 s
    B   6.4 s
    C   9.6 s
    D   12.8 s

12. Which particle causes fission of a uranium-235 nucleus?

    A   proton
    B   electron
    C   neutron
    D   alpha

13. When an atom is ionised which particle does it lose?

    A   neutron
    B   proton
    C   electron
    D   alpha

14. In a paper mill a radioisotope is used to monitor the thickness of the paper. Which isotope would be best?

    A   alpha emitter, short half-life
    B   beta emitter, short half-life
    C   beta emitter, long half-life
    D   alpha emitter, long half-life

15. In medicine a radioisotope is used to kill cancer cells. Which isotope would be best?

    A   alpha emitter, short half-life
    B   gamma emitter, short half-life
    C   gamma emitter, long half-life
    D   alpha emitter, long half-life

16. A radioisotope is used to monitor leaks from a water pipe. Which isotope would be best?

    A   gamma emitter, short half-life
    B   beta emitter, short half-life
    C   beta emitter, long half-life
    D   gamma emitter, long half-life

17. In medicine a radioisotope is used to monitor blood flow in a patient. Which isotope would be best?

    A   alpha emitter, short half-life
    B   gamma emitter, short half-life
    C   gamma emitter, long half-life
    D   alpha emitter, long half-life

Language focus — Glossary

a.c. Alternating current; constantly changing direction in the circuit

absorption Taking in

acceleration Increase in velocity per second

accurate Close to the true value

adjacent Next to

air resistance The force that opposes motion when an object moves through air

alpha particle A helium nucleus, emitted from a radioactive nucleus

ampere Unit of current

amplitude Height of a wave, measured from the undisturbed position to the maximum displacement

apparatus Equipment for an experiment

atom Particle from which all elements are made

attract Pull together

attract Push apart

background radiation The radiation due to natural and man-made sources that is all around us

balance Used to measure the mass of an object

barometer Used to measure atmospheric pressure

battery A group of cells

beta particle A fast-moving electron emitted from a radioactive nucleus

boil Change from a liquid to a gas at a constant temperature

Bunsen burner A gas burner used to heat in the laboratory

cell A source of chemical energy that causes current to flow around a circuit

centre of mass The point at which all the mass of the object can be considered to act

centripetal force The force that acts into the centre of a circle when an object is moving in a circle

charge Fundamental quantity that can be positive or negative

compress Squash into a smaller length or space

condense Change from a gas to a liquid

conduction Passing on of heat or electrical current by the particles in the object

conductor Able to easily pass on heat or electrical current

conservation Keeping the same before and after something has occurred

contract Become smaller

control Keep the same

convection The rising and falling of fluid as it is heated and becomes less dense and then cools and becomes more dense

convex Thicker in the middle than at the edges

coulomb The unit of charge

critical angle The angle at which total internal reflection occurs as light goes from a more to a less dense medium

current Charge passing a point each second in a circuit

d.c. Direct current; charge flow in one direction in a circuit

data Collected measurements

deflect Change direction

density The mass per unit volume of a material

dependent variable The variable that you measure in an experiment

diffraction The spreading out of a wave as it goes through a gap of suitable size

diode An electrical component that has very high resistance in one direction and therefore only lets current flow one way in a circuit

directly proportional When the independent variable is doubled, the dependent variable doubles

dispersion The splitting up of white light into a spectrum as it passes through a glass prism

displacement Distance in a certain direction from a position

distance Length an object has moved

domain Tiny magnet from which a ferromagnetic material is made

e.m.f. Electromotive force; the maximum potential difference a cell can supply

earth The ground, to which charge will flow if there is a conducting path

echo The reflection of a sound wave

efficiency The percentage of useful energy compared to the input energy

electromagnet An object that produces a magnetic field when a current flows through it

electromagnetic induction The production of a potential difference by cutting magnetic field lines with a conductor

electromagnetic spectrum The family of waves of varying frequency that have the same speed in a vacuum

Language focus — Glossary

electron A negatively charged particle of negligible mass

electrostatic The force due to a charged particle

emission Giving out

endoscope A medical instrument, which uses the principle of total internal reflection to look inside the body

equilibrium When no net force acts

evaporate Change from a liquid to a gas at a range of temperatures

expand Increase in size

extension Increase in length

ferromagnetic Having magnetic properties associated with iron

field The area around a charged particle or magnet where another charged particle or magnet experiences a force

filament lamp A bulb containing a piece of wire that glows when a current is passed through it

focal length The distance between the centre of a lens and the principal focus

force A push or a pull exerted on an object

freeze Change from a liquid to a solid

frequency The number of waves passing a point per second or the number of vibrations per second

friction The force acting between two surfaces that opposes the motion

fuse A safety feature in a circuit, which melts when the current exceeds a certain value

gamma ray An electromagnetic wave emitted from the nucleus of a radioactive element

generator Device that converts kinetic to electrical energy

gravitational field strength The gravitational force on 1 kg

half-life The average time taken for half of the nuclei of a radioisotope to decay

height Distance from the bottom to the top of an object

hypothesis Prediction about what will happen in an experiment based on scientific principles

image Where rays from an object meet in focus

incident Arriving/hitting

incompressible Not able to be squashed into a smaller volume

independent variable The variable that is changed in an experiment

insulator A material that does not conduct electricity

inversely proportional When the independent variable is doubled, the dependent variable halves

ion An atom that has lost or gained one or more electrons

ionising Having the ability to remove electrons from atoms

irregular Not having a particular geometric shape

isotope An atom of an element that has the same number of protons but a different number of neutrons

joule The unit of energy

kilogram The unit of mass

lens A shaped piece of glass which can focus light rays at a point

light-dependent resistor (LDR) An electrical component whose resistance decreases as the intensity of light increases

live The wire in a mains circuit that is connected to the alternating voltage supply

logic gate An electrical component whose inputs and output can have one of two states: high or low

longitudinal A wave in which the vibrations are parallel to the direction of wave travel

magnet An object around which there is a magnetic field

manometer A measuring instrument used to measure gas pressure

mass The amount of particles in an object

measuring cylinder Used to measure the volume of liquids

melt Change from a solid to a liquid

metre rule Used to measure the length of objects

moment Force multiplied by perpendicular distance from the pivot; the turning effect of a force

momentum Mass multiplied by velocity

motor A device used to change electrical energy to kinetic energy

negligible Almost zero

neutral Having no charge

neutron A particle of zero charge found in the nucleus of an atom

newton The unit of force

newton meter Used to measure force

non-renewable An energy resource that will eventually run out

Language focus — Glossary

normal A construction line lying at right angles to a boundary between two materials

nuclear fission The splitting of heavy nuclei into smaller nuclei with the emission of energy

nuclear fusion The joining of lighter nuclei into heavier nuclei with the emission of energy

nucleus The positively charged central part of an atom

observe Record what is happening

ohm The unit of resistance

parallel Two lines that have the same continuous distance between them. In a circuit, when components are connected across each other

particle A very small object, which cannot be seen with the naked eye

penetrating Able to pass though materials

pivot A point about which an object can turn

potential difference The work done on a coulomb of charge between two points in a circuit

power The energy transferred per second

precise When a measurement is taken and the repeats are very similar

prediction A statement about what is going to happen in an experiment, based on scientific knowledge and understanding

pressure Force per unit area

principal axis The imaginary line that passes through the centre of a lens

principal focus The point on the principal axis to which all rays converge after passing through a convex lens

probability The calculated chance that something may occur

proton A positively charged particle found in the nucleus of an atom

radiation Particles or waves that are emitted from the nuclei of radioactive elements

radioactive Having an unstable nucleus and emitting radioactive particles

radioisotope An isotope of an element that has unstable nuclei

radius The distance from the centre of a circle to its edge

rarefaction An area of a longitudinal wave where the particles are further apart than average

ray A thin beam of light

real An image that can be focused on a screen

reflection When an incident wave bounces off an object

refraction When an incident wave changes direction on passing into another medium

refractive index The ratio of speed of light in a vacuum to speed of light in a medium

relationship How one variable affects another

relay An electrical component that uses a small current to switch on a high current

reliable Predictable

renewable An energy resource that will never run out

repel Force apart

resistance Potential difference required to make one ampere of current flow

resistor An electrical component that reduces the current flowing in a circuit for the same voltage

resultant force The overall force on an object, taking into account all the forces that act

series Components that are connected into a circuit next to each other

solidify Change from a liquid to a solid

specific heat capacity The energy required to increase the temperature of 1 kg of substance by 1°C

specific latent heat The energy required to melt or vaporise 1 kg of substance

speed The distance travelled per second

stop clock Used to measure time

stretch Increase the length of an object

table A place to record data from an experiment

temperature The degree of 'hotness' of an object, related to the internal energy

thermal A type of energy to which all other types transform

thermistor An electrical component whose resistance decreases as temperature increases

thermometer Used to measure temperature

total internal reflection When light is incident in a denser medium at an angle greater than the critical angle and all of the light is reflected

tracer A radioisotope that is injected into the body and used to investigate processes in the body

transfer Move from one place to another

transformation Change into another type of energy

transformer A device used to step up and step down alternating voltage

Language focus — Glossary

transverse A wave whose vibrations are at right angles to the direction of wave travel

unreliable Not predictable

vacuum No particles present

variable resistor A resistor whose resistance can be changed by moving a sliding contact

velocity Speed in a certain direction

Vernier callipers A measuring instrument used to measure small distances

vibration A back and forth motion

virtual An image that cannot be focused onto a screen

volt The unit of potential difference

volume The space occupied by a solid, liquid, or gas

wavefront A line along the wave where all the particles are at the same point in their vibration

wavelength The shortest distance between two particles that are at the same point in their vibration

weight The force on an object due to gravity

work Energy transferred due to the action of a force over a distance

Language focus — Key word exercises

Exercise 1

1. Name the three forces that act on an object falling through water.

   ....

2. Name the three forces that act at a distance away from the object.

   ....

3. Name three types of stored energy.

   ....

4. State the quantities that have the following units:

   N .... Nm .... m/s ....

   m .... m/s^2 .... Pa ....

5. Name three renewable energy resources.

   ....

6. Name two energy resources that do not originate from the Sun.

   ....

7. Name the instrument that is used to measure gas pressure. ....

8. State the force that opposes the motion of two surfaces rubbing together. ....

9. Name the force that opposes the motion of an object moving through air. ....

10. Name the quantity which is equal to mass divided by volume, and describes how close particles are packed in a substance. ....

11. Name an instrument used to measure length. ....

12. Name the instrument used to measure atmospheric pressure. ....

Supplement

13. When the force on an object is doubled, the acceleration doubles. What is the name given to this relationship?

    ....

14. What is the name given to a change in momentum? ....

15. State the Principle of Conservation of Momentum.

    ....

    ....

Language focus — Key word exercises

Exercise 2

1. Solid, liquid, and gas are all ....
2. How do the particles move in a solid? ....
3. Which of solids, liquids, and gases are compressible? ....
4. How do particles in a gas move? ....
5. What is the name given to the process by which liquids change to gases at a range of temperatures?

   ....
6. What is the name given to the process by which liquids change to gases at a constant temperature?

   ....
7. When a solid is heated the particles vibrate more and move further apart.

   This is called ....
8. The expansion of a bimetallic strip can be used in a ....
9. Which of solids, liquids, and gases expands the most on heating? ....
10. The electrical component whose resistance changes with temperature. ....
11. The energy associated with the vibration of particles in an object. ....
12. The energy required to melt a solid is called the ....
13. The reverse of evaporation is ....
14. The best conductors of heat are all ....
15. When water is heated it .... because the particles increase their

    ....
16. What type of radiation is emitted from a hot surface? ....
17. When warm air rises, a .... is set up.

Supplement

18. J/(kg °C) is the unit of ....
19. When the pressure on a gas is doubled the volume is halved. What is the name of this relationship?

    ....
20. Absolute zero is 0 on which scale of temperature? ....

Language focus   **Key word exercises**

Exercise 3

1. What is the name for the distance between adjacent particles at the same point in their vibration in a wave?

   ....

2. What is the distance between the centre and the maximum displacement in a wave? ....

3. What is the number of waves passing a point per second called? ....

4. Which quantities have the following units?

   Hz .... m/s ....

5. What is the name for a line along the wave where all of the particles are at the same point in their vibration?

   ....

6. What is the name for a wave changing direction as it travels into a denser medium? ....

7. Which two angles are the same when light reflects from a mirror? ....

8. Which type of image can be focused on a screen? ....

9. What us the name for a reflected sound wave? ....

10. Give three qualities that electromagnetic waves have in common.

    ....

    ....

    ....

11. If a wave approaches a gap of similar size to its wavelength, what phenomenon occurs? ....

12. When light passes through a triangular prism, it may split up into its component colours. What is this effect called?

    ....

13. State three qualities of the image formed in a plane mirror.

    ....

    ....

    ....

14. Name the three sections of the electromagnetic spectrum that are ionising.

    ....

15. When a ray of light is incident in glass on a glass–air boundary at an angle greater than the critical angle, what effect occurs? ....

Language focus — Key word exercises

Exercise 4

1. State the quantities that have the following units:

   V .... A .... C ....

   J .... W .... Ω ....

2. What is the name given to the area around a charge inside which another charge experiences a force?

   ....

3. What is the effect of a magnetic north pole on another north pole? ....

4. Name three factors that affect the strength of an electromagnet.

   ....

   ....

   ....

5. Name a component that transforms electrical energy to kinetic energy. ....

6. Name a component that uses a small current to switch on a larger current. ....

7. Name a component whose resistance changes with light intensity. ....

8. Name a component whose resistance increases with increasing potential difference.

   ....

9. When the cross-sectional area of a wire doubles, the resistance halves. What is the name for this relationship?

   ....

10. Which part of an electrical plug protects people from fires? ....

11. Which part of an electrical plug protects people from electric shocks? ....

12. When a magnet is moved into a coil a potential difference is ....

13. Name the device which is used to step up alternating potential difference. ....

Supplement

14. In Fleming's left-hand rule, what is represented by the thumb? ....

15. Which part of a d.c. motor allows the current to reverse as the coil spins? ....

16. In the a.c. generator, which parts allow the coil to keep turning? ....

17. Name the component that can be used to convert a.c. to d.c. ....

Language focus — Key word exercises

Exercise 5

1. Name the particles in the nucleus of an atom. ....
2. Name the particles that orbit the nucleus. ....
3. If the atom is neutral, the numbers of which particles in an atom must be equal?
   ....
4. What is the name given to the total number of particles in the nucleus? ....
5. In the discovery of the nucleus, which type of radioactive particles was fired at a thin gold foil? ....
6. What is the name of the process where a neutron causes a nucleus to split into two smaller nuclei and some neutrons? ....
7. What is the name of the process where smaller nuclei join to form larger nuclei? ....
8. Which type of nuclear radiation belongs to the electromagnetic spectrum? ....
9. Which type of nuclear radiation can be absorbed by a sheet of paper? ....
10. What is the name given to the time taken for half of the radioactive nuclei to decay? ....
11. Name a device used to detect radiation. ....
12. What is the name given to the process by which radioactive particles remove electrons from atoms?
    ....
13. What is the name of the device that radiation workers wear to ensure they do not receive too high a dose?
    ....

Supplement

14. Which type of radiation is undeflected by a magnetic field? ....
15. Which type of radiation is used to monitor the thickness of paper during manufacture? ....
16. In which procedure is a gamma emitter injected into a patient, and the emitted gamma radiation is monitored?
    ....
17. Cobalt-60 is a radioactive source that emits gamma radiation. If this gamma radiation is focused into the patient, what benefit is produced? ....
18. Which type of radiation is attracted to a positively charged metal plate? ....
19. In Rutherford's alpha scattering experiment, some alpha particles were deflected straight back. What does this tell us about the nucleus? ....
20. Why was Rutherford's experiment carried out in a vacuum? ....

Revision tips

Start early!

Revise little and often and test yourself. Trying to cram everything at the last minute is a mistake.

Use active revision techniques

Reading your notes is a very ineffective way of revising, as is going through them with a highlighter pen just highlighting key words.

Try reading one topic area from your notes and writing ten questions on that topic. Write the answers on another piece of paper. In a few days, weeks or months, you can use the questions to test yourself and check your answers.

Drawing mind maps will allow you to reorganise information and make links across a topic. Using diagrams and colour coding to link ideas will help you to make connections.

Summarising your notes onto flashcards will help you to take them with you when you are away from your books. This means you can test yourself when you have a spare ten minutes.

Ask another IGCSE student to test you on a particular topic. They may include questions that you had not thought of.

Above all, practise answering examination questions. Read the mark scheme carefully and learn from your mistakes.

Don't avoid difficult topics!

It is tempting to begin revising the areas of the syllabus on which you are most confident but it is better to tackle the most challenging topics first. Even though you may not feel that you fully understand after the first time you have revised the work, some ideas will begin to make sense. This means that each time you revisit the topic it will become easier to understand.

Answering examination questions

Read the question carefully, looking for cues that tell you how much detail is required in the answer

For example, the question 'Describe in terms of particles the mechanism of heat conduction in a metal', which is worth **two** marks, requires you to make **two** valid points and include details of the particles involved.

The word '**describe**' means that you have to say **what** is happening.

For example, the question 'Explain in terms of the movement of gas molecules why the pressure of a fixed mass of gas increases as its volume decreases', which is worth **two** marks, requires you to make **two** valid points and include details of the movement of the gas molecules.

The word '**explain**' or phrase '**give a reason**' means that you have to say **why** something is happening.

The first line of a question gives introductory information that you may need to refer to throughout the question. Underlining key information, such as values of variables, can help.

Take care to answer the question asked. Physics problems are not solved by writing down everything you know about a topic or answering a similar question that you have practised before. You must apply your knowledge to the specific example presented.

Revision tips

Be precise in giving your answers

Avoid giving vague answers or using general terms. Take care to use specific physics key words such as energy and force in the correct context. Do not repeat statements from or reword the question, as this does not carry any marks.

Show full working in calculations

Rearrange the equation to make the variable you are trying to find the subject (so that it is on the left-hand side of the equation), put in the numbers (with the correct units) and then use your calculator to find the answer. Give your answer to two or three significant figures with the correct unit.

You should be competent at using your calculator. Do not buy a different model just before the exam!

Check your answer and decide if it seems sensible – did you expect something similar?

Answering multiple choice questions

Take great care to mark your answer in the correct place. Do not worry if you put the same answer several times in a row. There is no particular distribution of As, Bs, Cs, or Ds. If you are stuck on an answer move on and make a careful note of which one you have left out. Take even more care to make sure the next answer goes in the correct place.

Revision checklists

Tick to indicate when you have completed your revision mind maps and /or flash cards for each topic and when you have practised questions of exam standard. Shaded boxes indicate the material is supplement only.

Topic/skill area	Made revision mind maps / flashcards?	Practised exam questions?
Taking measurements		
Distance, speed, time, and acceleration		
Mass and weight		
Density		
Forces and their effects		
Moments and equilibrium		
Centre of mass		
Principle of Conservation of Momentum		
Energy types and the Principle of Conservation of Energy		
Energy resources		
Work and power		
Kinetic theory		
Evaporation		
Pressure		
Thermal expansion		
Measurement of temperature		
Thermal capacity and specific heat capacity		
Melting, boiling, and latent heat		
Conduction		
Convection		
Radiation		
Applications of conduction, convection, and radiation		
Wave properties		
Reflection		
Refraction		

Revision checklists

Topic/skill area	Made revision mind maps / flashcards?	Practised exam questions?
Lenses		
Dispersion		
The electromagnetic spectrum		
Sound		
Magnetism		
Electric charges		
Current		
E.m.f. and potential difference		
Resistance		
Energy and power in electric circuits		
Circuit diagrams and the use of components in circuits		
Series and parallel circuits		
Digital electronics		
The dangers of electricity		
Electromagnetic induction		
The a.c. generator		
The transformer		
The magnetic field due to a current		
Force on a current-carrying conductor		
The d.c. motor		
The structure of the atom		
Detection of radiation		
The three types of radiation		
Radioactive decay		
Half-life		
Radiation safety		

Mind maps

Forces and motion revision mind map for you to complete

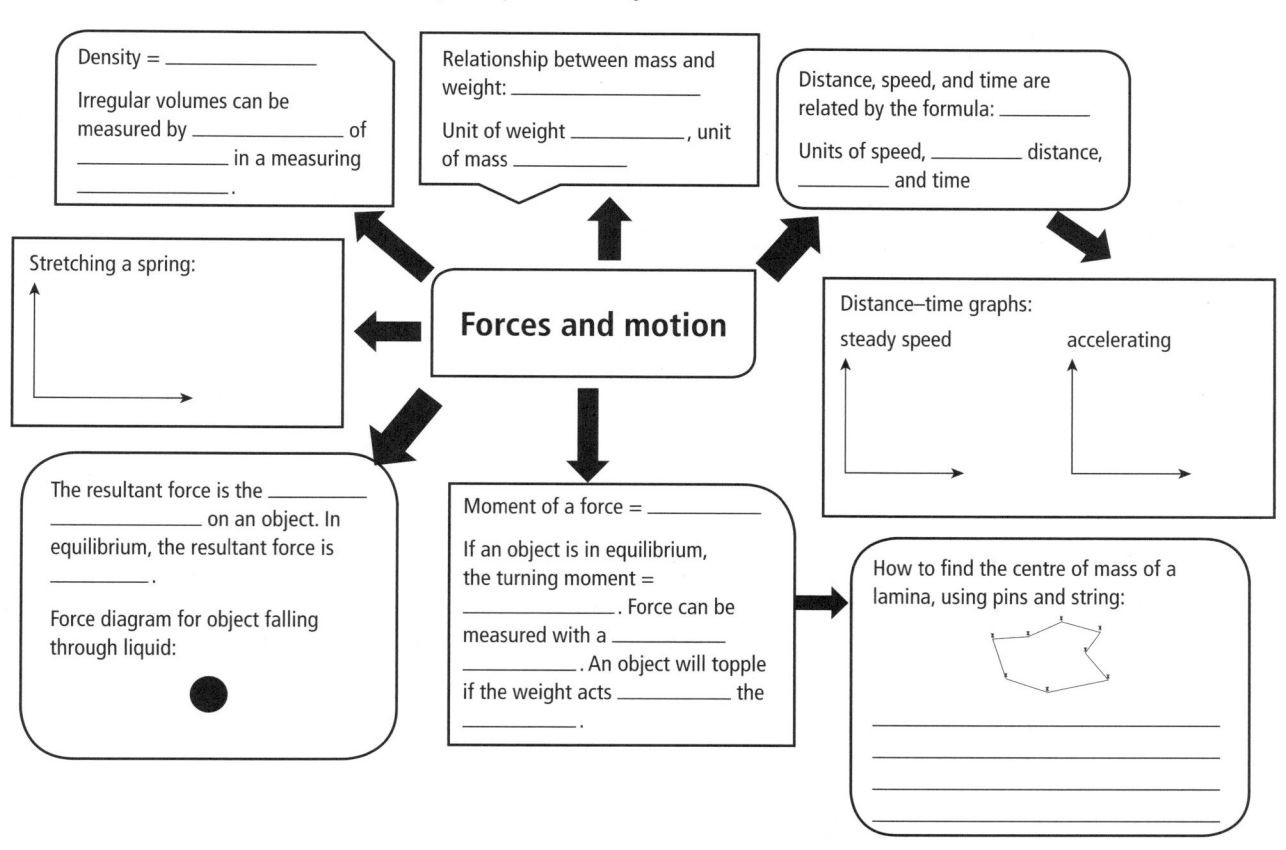

- Density = _____
  Irregular volumes can be measured by _____ of _____ in a measuring _____.

- Relationship between mass and weight: _____
  Unit of weight _____, unit of mass _____

- Distance, speed, and time are related by the formula: _____
  Units of speed, _____ distance, _____ and time

- Stretching a spring:

- **Forces and motion**

- Distance–time graphs:
  steady speed   accelerating

- The resultant force is the _____ _____ on an object. In equilibrium, the resultant force is _____.
  Force diagram for object falling through liquid:

- Moment of a force = _____
  If an object is in equilibrium, the turning moment = _____. Force can be measured with a _____ _____. An object will topple if the weight acts _____ the _____.

- How to find the centre of mass of a lamina, using pins and string:

Work, energy, and power revision mind map for you to complete

- If the useful output energy is 60 J and input energy is 80 J, the efficiency can be calculated:
  Efficiency = ─────── × 100%
             = _____ %

- Energy types:

- Renewable energy resources will not _____.
  Examples that originate from the Sun:
  _____
  Examples that do not originate from the Sun:
  _____

- **Work, energy, and power**

- Magic triangles for calculating work done and power:

- Energy transformation in a motor:
  electrical ⟶ _____ and
  wasted _____ and _____
  Energy transformation in a television:
  electrical ⟶ _____ and
  _____ and wasted _____

- Transformation of gravitational potential energy to kinetic energy:
  A boy of mass 50 kg climbing a hill of height 3 m gains GPE:
  50 × _____ × _____ = _____ J
  If he slides down the hill and his GPE is transformed to KE he gains speed:
  $v = \sqrt{\frac{2KE}{m}} = \sqrt{(\text{_____}/50)} = \text{_____}$ m/s

131

Mind maps

Thermal physics mind map for you to complete

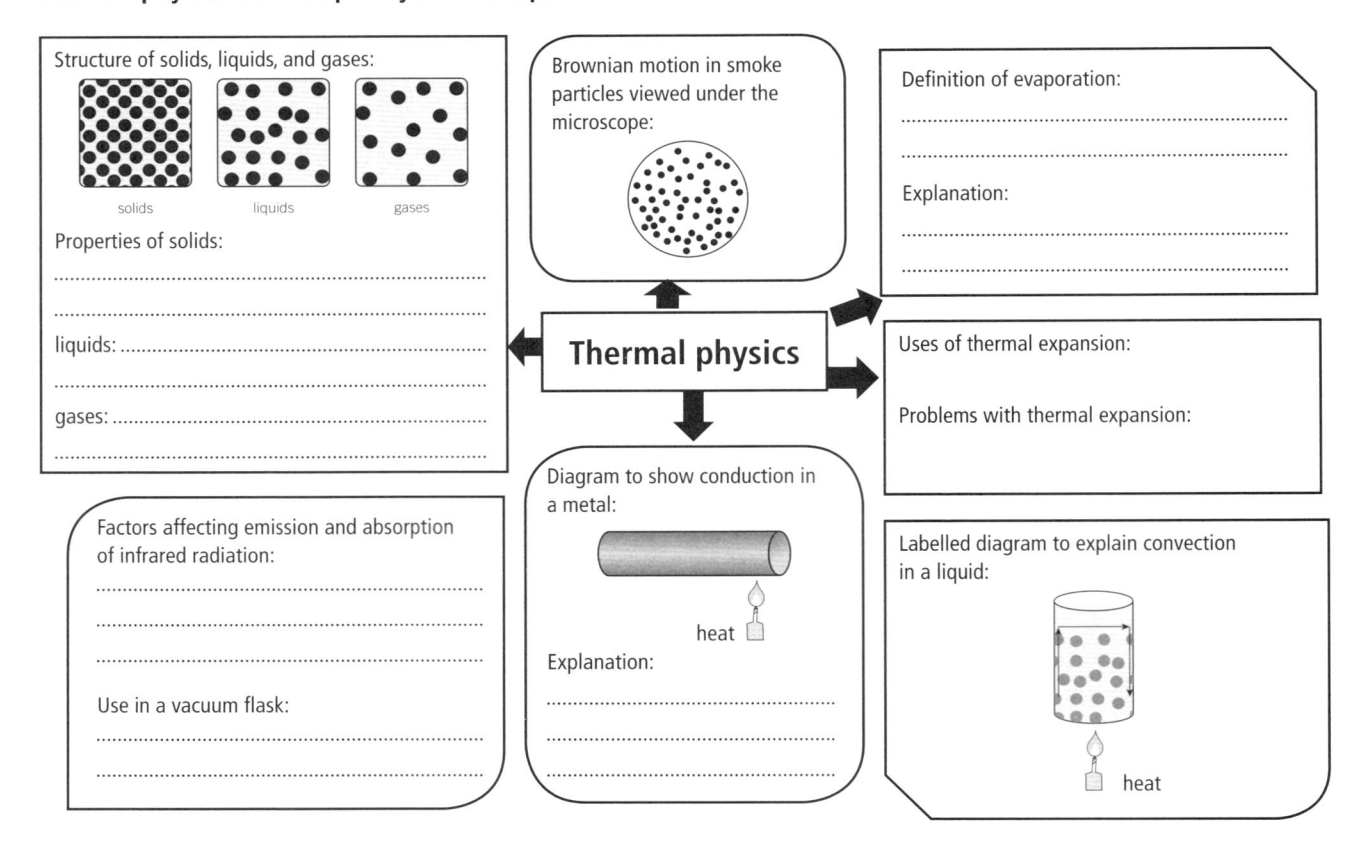

Structure of solids, liquids, and gases:

solids   liquids   gases

Properties of solids:
....
....
liquids: ....
....
gases: ....
....

Brownian motion in smoke particles viewed under the microscope:

Definition of evaporation:
....
....
Explanation:
....
....

→ **Thermal physics** ←

Uses of thermal expansion:

Problems with thermal expansion:

Factors affecting emission and absorption of infrared radiation:
....
....
....
Use in a vacuum flask:
....
....
....

Diagram to show conduction in a metal:

heat

Explanation:
....
....
....

Labelled diagram to explain convection in a liquid:

heat

A mind map on waves for you to complete

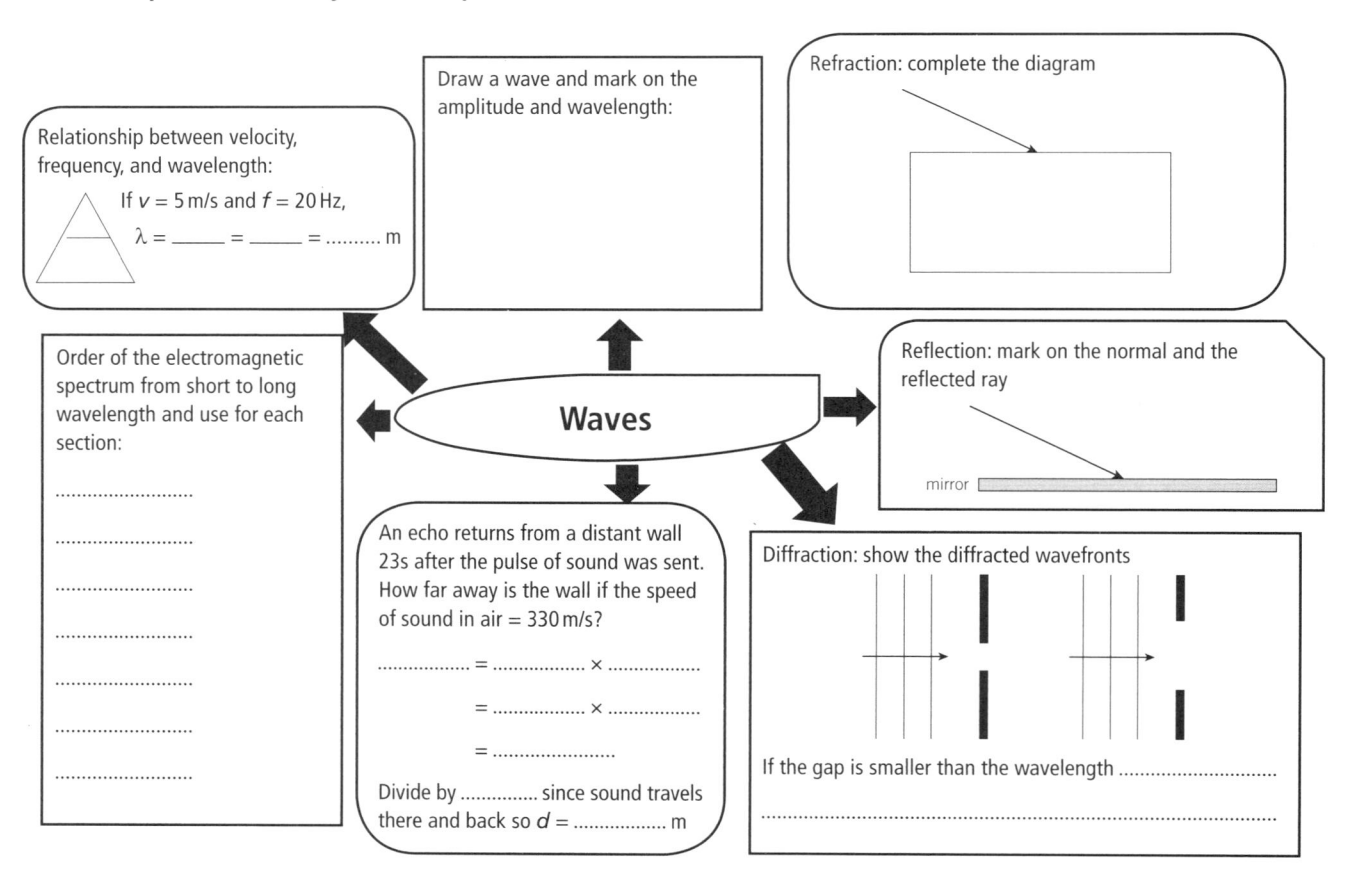

Relationship between velocity, frequency, and wavelength:

If $v = 5$ m/s and $f = 20$ Hz,

$\lambda = $ _____ = _____ = ... m

Draw a wave and mark on the amplitude and wavelength:

Refraction: complete the diagram

Order of the electromagnetic spectrum from short to long wavelength and use for each section:
....
....
....
....
....
....

→ **Waves** ←

Reflection: mark on the normal and the reflected ray

mirror

An echo returns from a distant wall 23 s after the pulse of sound was sent. How far away is the wall if the speed of sound in air = 330 m/s?

.... = ... × ....

= .... × ....

= ....

Divide by .... since sound travels there and back so $d = $ .... m

Diffraction: show the diffracted wavefronts

If the gap is smaller than the wavelength ....
....

132

Mind maps

A mind map on electricity for you to complete

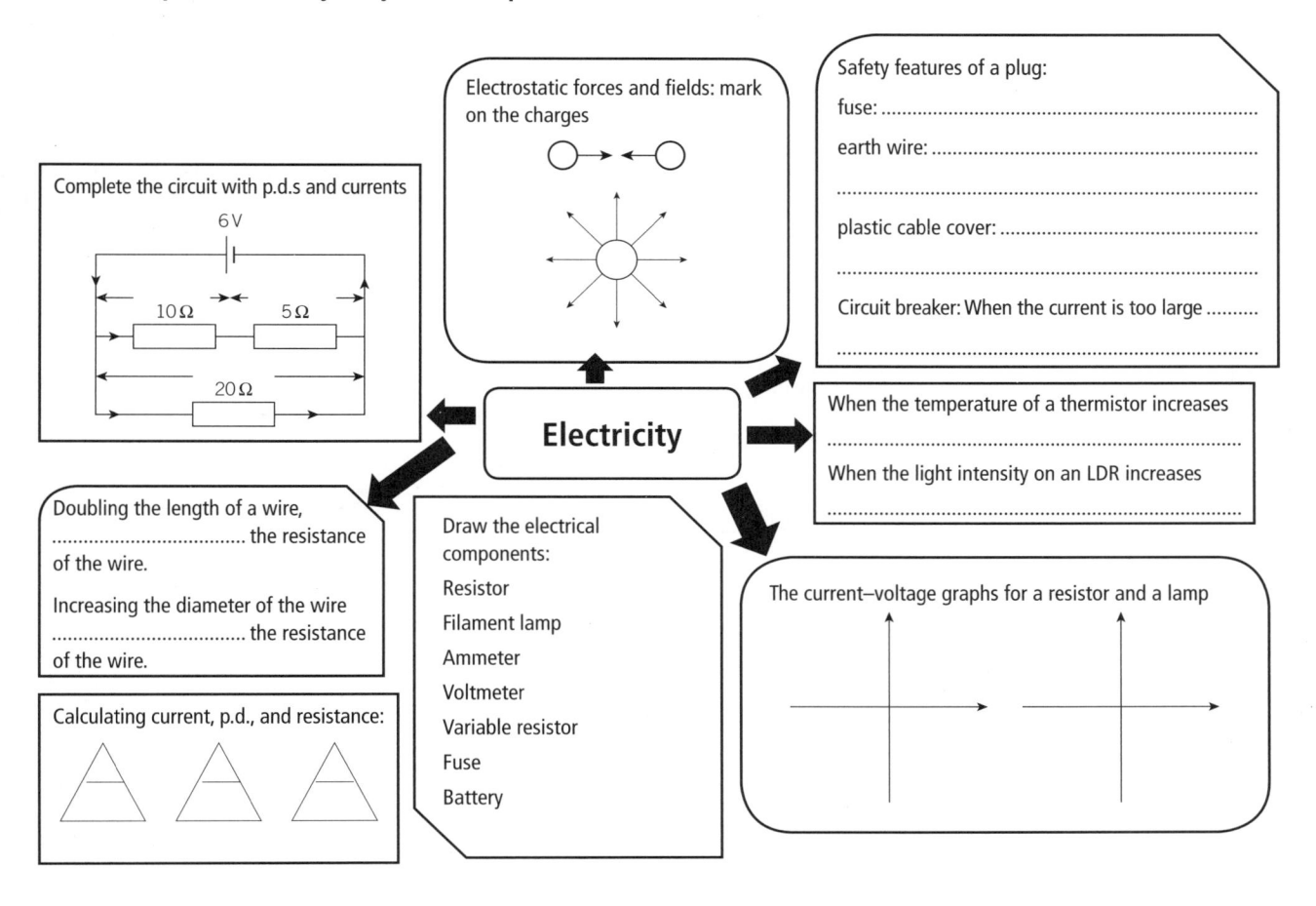

Complete the circuit with p.d.s and currents
(Circuit: 6V battery, 10Ω and 5Ω resistors in series on top branch, 20Ω resistor on bottom branch)

Electrostatic forces and fields: mark on the charges

Safety features of a plug:
fuse: ....
earth wire: ....
....
plastic cable cover: ....
....
Circuit breaker: When the current is too large ....
....

Electricity

When the temperature of a thermistor increases
....
When the light intensity on an LDR increases
....

Doubling the length of a wire, ... the resistance of the wire.
Increasing the diameter of the wire .... the resistance of the wire.

Calculating current, p.d., and resistance:
(three triangles)

Draw the electrical components:
Resistor
Filament lamp
Ammeter
Voltmeter
Variable resistor
Fuse
Battery

The current–voltage graphs for a resistor and a lamp
(two axes)

A mind map on magnetism for you to complete

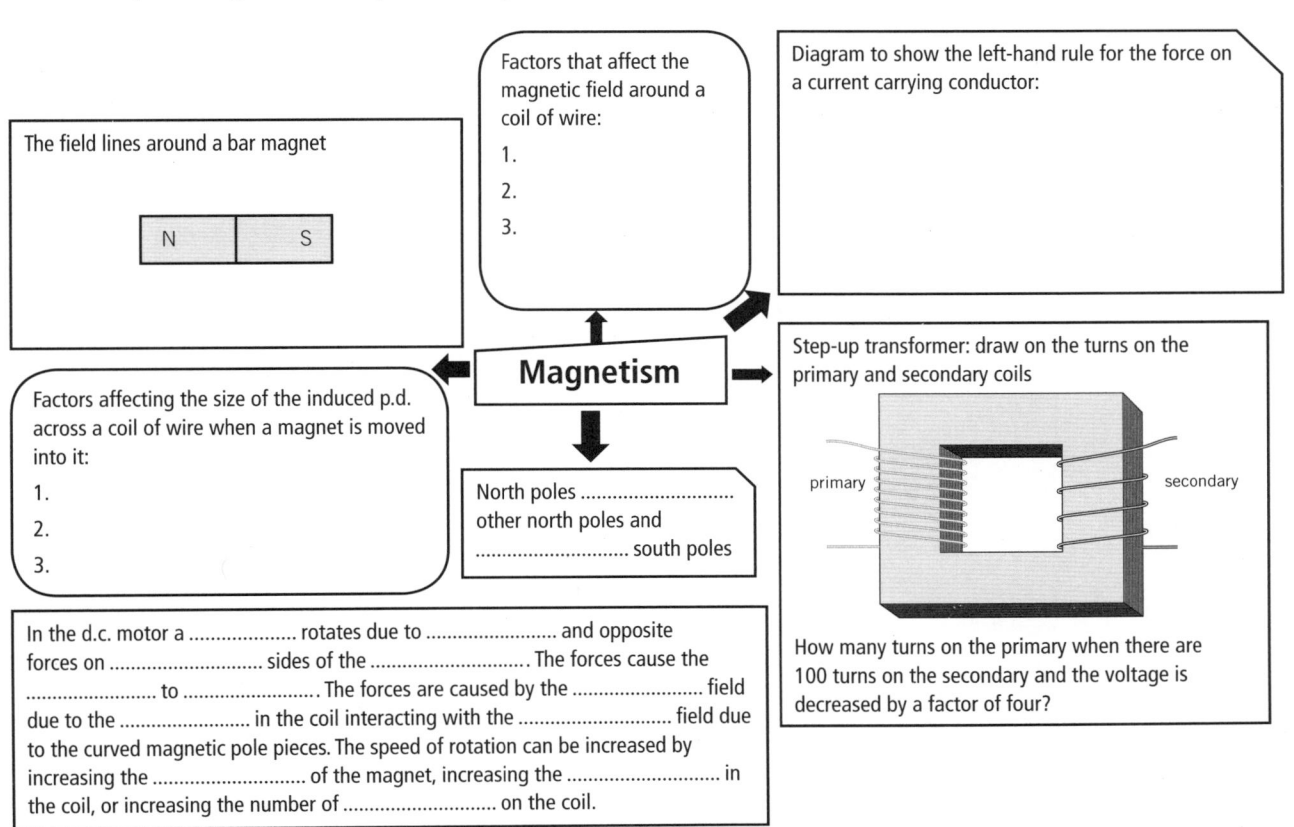

The field lines around a bar magnet
(N S bar magnet)

Factors that affect the magnetic field around a coil of wire:
1.
2.
3.

Diagram to show the left-hand rule for the force on a current carrying conductor:

Magnetism

Factors affecting the size of the induced p.d. across a coil of wire when a magnet is moved into it:
1.
2.
3.

North poles .... other north poles and .... south poles

Step-up transformer: draw on the turns on the primary and secondary coils
(transformer diagram labelled primary and secondary)

How many turns on the primary when there are 100 turns on the secondary and the voltage is decreased by a factor of four?

In the d.c. motor a .... rotates due to .... and opposite forces on .... sides of the .... The forces cause the .... to .... The forces are caused by the .... field due to the .... in the coil interacting with the .... field due to the curved magnetic pole pieces. The speed of rotation can be increased by increasing the .... of the magnet, increasing the .... in the coil, or increasing the number of .... on the coil.

133

Mind maps

A mind map for you to complete on atomic physics

The structure of the atom:

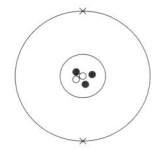

Mark on 4 electrons, the correct number of protons for a neutral atom, and a suitable number of neutrons.

Relative mass and charge of subatomic particles

	relative mass	charge
proton		
neutron		
electron		

Number of protons ....

Number of protons + neutrons ....

Discovery of the nucleus:

A thin .... foil was bombarded with .... particles. Most of the .... particles went straight .... without being .... which shows that the atom is mainly .... Some of the .... particles were deflected through angles of up to .... degrees which shows that the atom contains a concentrated .... charge in a small ....

Atomic physics

Half-life is ....
....

Sketch the graph of activity v time for a radioisotope

If the half-life of a radioisotope is 3 hours, how long does it take to fall to 1/8th of its original activity?

Radiation

	what?	ionising?	penetrating?
alpha			
beta			
gamma			

Alpha radiation is absorbed by a few centimetres of .... or a sheet of ....

Beta radiation is absorbed by a few millimetres of ....

Gamma radiation is absorbed by thick .... or concrete.

Nuclear fission: label the diagram

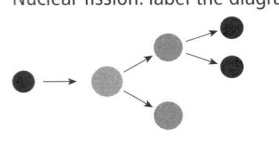

Nuclear fission is used in nuclear ....

Nuclear fusion: label the diagram

Nuclear fusion occurs in the ....

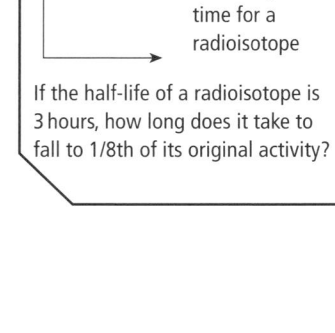

Practical skills

A laboratory report should include:

1. A risk assessment for the practical, including the possible hazards and how the risks that these hazards cause can be controlled.

e.g. a hot bulb is a hazard because it may burn you, but you can control the risk by switching off the circuit in between readings and not touching the bulb.

2. Details of any preliminary experiments you carry out to work out the range and increments of the independent variable.

e.g. in an experiment to find the resistance of an LDR at different distances from a bulb, you may choose to increase the distance of the LDR from the bulb in 2 cm increments up to a range of 20 cm. If this does not produce great enough changes in resistance, you should record the results and then state any changes to your plan to include a greater range of distances and larger increments.

3. A brief description of the experimental procedure, including labelled diagrams of the apparatus.

The diagrams should be drawn in pencil in 2-D. They need not be complicated but should have all relevant apparatus labelled so that the set-up is clear.

It should be clear how your experimental technique will lead to you to obtain reliable, accurate results and how you have controlled your control variables.

4. Tables of measurements with headings and units. All measurements must be recorded and not just the derived values.

e.g. if you are finding resistance by measuring potential difference and current, you must record all the values of potential difference and current.

5. Appropriate calculations using your data.

If you believe the results are not correct, you should not throw them away or scribble them out. Any anomalies (results that do not fit into the pattern) should be circled and remain in your tables. When you repeat your experiment and obtain results that are in agreement with all your other results, you should not include the anomalous data in calculating your averages.

6. A graph of the results of the experiment, including the following:

- An appropriate scale on each axis, so that the plotted points fill as much of the graph paper as possible (at least half).
- Axes labelled with both the variables and their units.
- Points plotted as small pencil crosses.
- A straight line of best fit drawn with a ruler, or a smooth curved line of best fit drawn freehand.

    (A line of best fit does not have to pass through every point but there should be an equal number of points above and below the line.)

- A title that clearly states which variables are being investigated and which variables are controlled.

    (e.g. "A graph to show how the potential difference across a bulb affects the current flowing through the bulb".)

Practical skills

> **7.** A conclusion reached using the data collected in the experiment.

You should be able to state the trend shown by the graph. If you note a particular relationship (e.g. direct proportion where the graph is a straight line through the origin) then you should state this. Otherwise give the general trend. Back up your statement with evidence from the graph.

> **8.** Evaluate your results for reliability and accuracy.

Consider the distribution of points about your best fit line. If the points are all distributed on or close to the line of best fit, this provides more reliable evidence for your conclusion.

Consider variables that were difficult to measure accurately e.g. the time period of a fast-swinging pendulum or a very small distance, and explain how you used multiple measurements to increase accuracy.

Example laboratory report for you to complete

An experiment to find the effect of different coloured surfaces on heat loss from a beaker of water.

Apparatus

Three 250 ml beakers, measuring cylinder, polystyrene mat, thermometer, black paper, silver paper, white paper, stop clock, 3 cardboard lids with a hole in the centre, kettle

Diagram

Complete the labels on the diagram

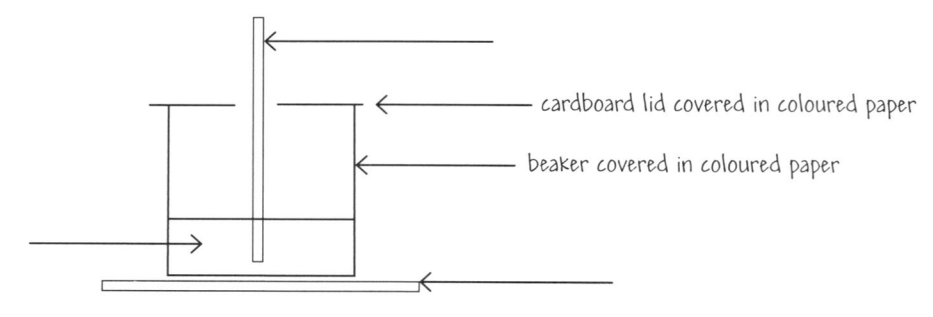

- cardboard lid covered in coloured paper
- beaker covered in coloured paper

Risk assessment

Complete the risk assessment by adding in the control measures

Hazard	Risk	Control
Glass beakers	If dropped the glass beaker could cut you.	Keep the apparatus away from the edge of the bench.
kettle	The sides of the kettle will become hot and could burn you.	
Hot water	If spilled, the hot water could scald you.	

Practical skills

Plan

Complete the plan of how to do the experiment

Take three glass beakers and wrap one with ...., one with ...., and one with .... Cover three cardboard lids with ...., ..., and .... paper and cut a hole in the middle of each of the lids for the .... Place the first beaker on the polystyrene mat. Measure out 250 ml of boiling water from a kettle using a .... Pour the hot water into the .... Place the lid on the beaker and put the .... through the hole in the lid into the beaker. When the temperature on the .... reads 80 °C, start the .... After 20 minutes take the .... again with the .... Repeat the experiment for each beaker, ensuring that 250 ml of water is used each time, that the starting temperature is .... each time, and that .... Repeat the whole experiment again for each beaker.

Results table

Complete the table

beaker	start temperature / °C	end temperature / °C	temperature change / °C	average temperature change / °C
black	80	45	35	37
	80	42	38	
white	80	54		
	80	49		
silver	80	58		
	80	56		

Conclusion

Complete the sentences for the conclusion

The beaker with the black paper showed the .... change in .... after 20 .... The beaker with the silver paper showed the .... change in .... after .... The .... paper caused the .... heat loss from the beaker and the silver paper the .... heat loss.

137

Practical skills

Evaluation

Answer the following questions to evaluate the experiment

Were there any results in the table that did not fit in with the pattern (anomalous results)?

Was there enough data in the table to decide which coloured beaker lost the most heat?

Which measurements were most difficult to take (e.g. ones where you would need to measure two things at the same time)?

How could the experiment be improved to obtain more accurate results?

Practical skills

Preparing for the practical or written alternative to the practical

Each of the practical skill areas 1 to 8 may be tested using specific examples from the syllabus, such as:

- cooling and heating
- springs and balances
- electric circuits
- reflection and refraction
- timing motion or oscillations
- measuring quantities such as length, volume, or force.

You may be required to:

- Use simple apparatus, in situations where the method may not be familiar to you.
- Select the most appropriate apparatus or method for a task and justify your choice.
- Explain how the apparatus can be used to find a derived quantity (e.g. resistance) or test the relationship between two variables.

Examination-style questions on practical skills

1. a. Name the following measuring instruments:

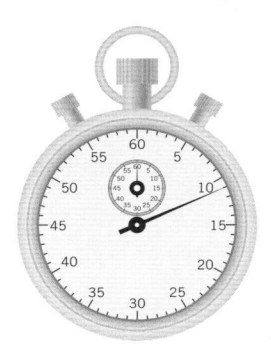

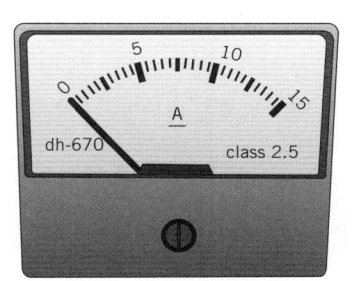

   .... [3]

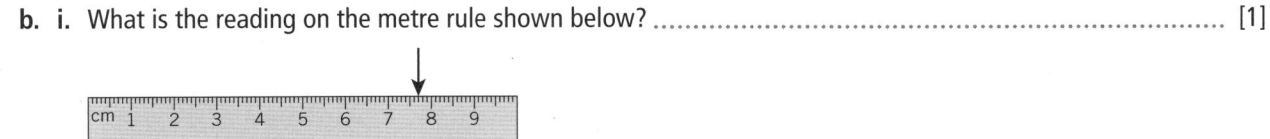

   b. i. What is the reading on the metre rule shown below? .... [1]

   ii. What is the reading on the Vernier callipers shown below? .... [1]

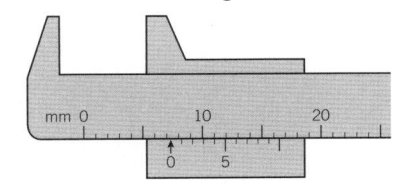

   iii. What is the reading on the micrometer shown below? .... [1]

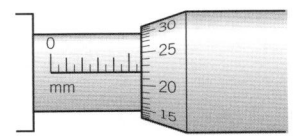

139

Practical skills

2. a. In the table below there are some measured quantities and suggested values. Circle the correct value for each quantity.

mass of a pencil	5 g
	5 mg
	5 kg
surface area of a test tube	0.75 cm²
	75 cm²
	750 cm²
thickness of a protractor	0.2 mm
	2 mm
	20 mm
weight of an apple	0.010 N
	0.10 N
	1.0 N
volume of a metre rule	76 mm³
	76 cm³
	76 m³

[5]

b. An IGCSE student finds by experiment the power of a bulb by measuring the current and potential difference for the bulb and using the formula:

$$P = IV$$

Suggest one way in which the student can obtain a reliable value for power and explain how this method would lead to a reliable result.

....

....

.... [2]

c. The diagrams show the reading on the voltmeter and corresponding ammeter reading.

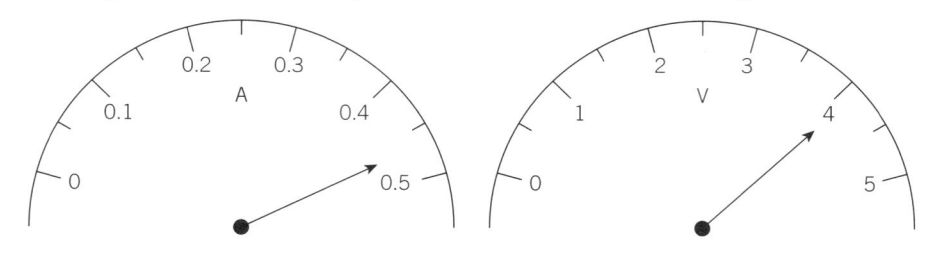

i. Write down the reading of current. .... [1]

ii. Write down the reading of potential difference. .... [1]

iii. Calculate the power of the bulb at this potential difference.

power = ... [2]

140

Practical skills

3. An IGCSE student is investigating the time period, *T*, of a simple pendulum, consisting of a small heavy bob on a string, displaced through a small angle and released.

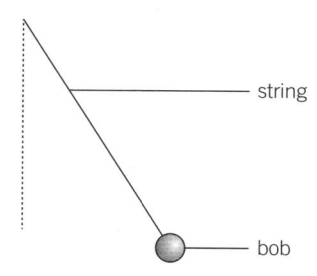

   a. Describe a method of accurately calculating the length of the pendulum, which is measured to the centre of mass of the bob.

   ....

   ....

   .... [3]

   The student measures the time for 10 oscillations, *t*, and obtains the following results:

l / m	t / s	T / s	$\frac{T}{l}$ / $\frac{s}{m}$
0.900	18.8		
0.600	15.4		
0.300	10.9		

   b. Calculate the values of *T* and enter them in the table. [2]

   c. Calculate the values of $\frac{T}{l}$ and enter them in the table. [2]

   d. The student suggests that *T* should be directly proportional to *l*. State and explain whether the results in the table support this statement.

   ....

   ....

   .... [2]

   e. The swings of the pendulum are short and so it is difficult to accurately record the time for 1 swing. How has the student included a technique to obtain an accurate value for *T*?

   ....

   .... [2]

Practical skills

4. A group of IGCSE students are carrying out an experiment to measure the refractive index of Perspex using a rectangular Perspex block.

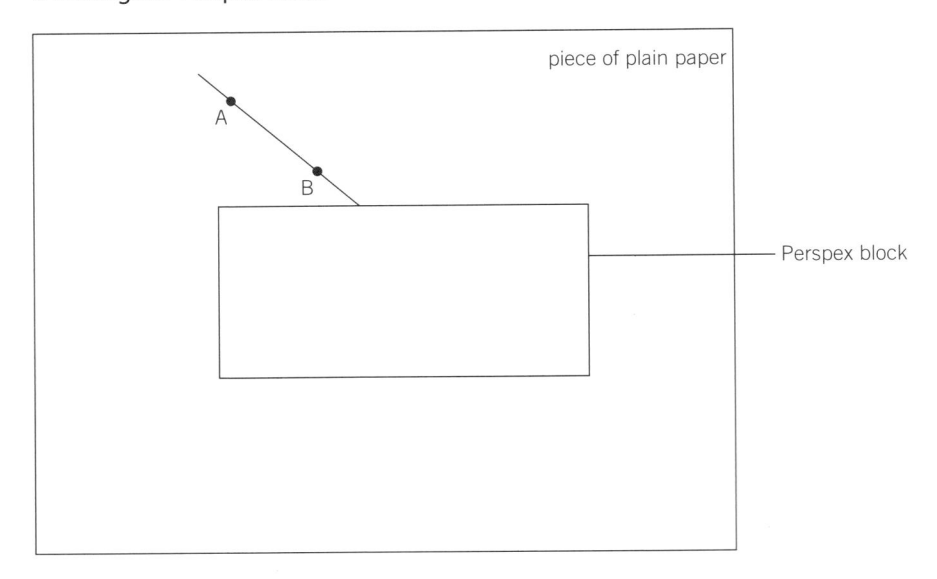

They place the block on a piece of plane paper on a pin board, draw a line along which to send a ray and mark the line in two positions, **A** and **B** by sticking a pin in each of these points.

a. On the diagram mark on the normal to the block where the line meets the block. [1]

b. Measure the angle of incidence. .... [1]

c. The angle of refraction inside the block is 24°. Mark on the refracted ray so that it is incident on the far side of the block. [2]

d. Mark on the ray that emerges from the far side of the block. [2]

e. Mark with the letters **C** and **D** two points where pins should be placed, so that when one of the students looks through the block, the four pins **A**, **B**, **C**, and **D** all appear to be in line with each other. [2]

Practical skills

5. An IGCSE student is investigating the cooling of water. She pours 150 ml of hot water into a 250 ml beaker and places a thermometer in the beaker.

   a. Name the measuring instrument used to measure the volume of water. .... [1]

   b. Record the following thermometer readings at the given times in the table. Complete the column headings.

   $t = 0$

   $t = 4$ minutes

   $t = 8$ minutes

   $t = 12$ minutes

t / s	
0	
4	
8	
12	

   c. i. What is the difference in temperature between $t = 0$ minutes and $t = 4$ minutes? .... [1]

   ii. What is the difference in temperature between $t = 4$ minutes and $t = 8$ minutes? .... [1]

   iii. What is the difference in temperature between $t = 8$ minutes and $t = 12$ minutes? .... [1]

   d. Use your answers to part **c.** to describe the rate of cooling with time.

   .... [1]

   Give a reason for your answer.

   ....

   .... [2]

   e. The student decides to repeat the experiment to check her results. Give two variables that must be controlled in the repeat test.

   ....

   .... [2]

143

Mathematics for physics

You need to practise and develop your maths skills in order to interpret and solve questions involving calculations.

You need to be able to:

1. Add, multiply, and divide to solve problems and rearrange equations.

2. Calculate averages by adding values together and dividing by the number of values.

e.g. the average of 2.0, 2.2, 2.2, 2.3 is $\frac{(2.0 + 2.2 + 2.2 + 2.3)}{4} = 2.2$

3. Calculate percentages by dividing a value by the total and multiplying by 100%.

e.g. as a percentage of 10 J, 2 J is $\left(\frac{2}{10}\right) \times 100\% = 20\%$

4. Use decimals and fractions in calculations.

e.g. $\frac{1}{4} = 0.25$

5. Understand ratios and convert to fractions and percentages.

e.g. a ratio of 1:4 means a fraction of $\frac{1}{5}$ or a percentage of 20%

6. Work out the reciprocal of a number by dividing 1 by the number.

e.g. the reciprocal of $7 = \frac{1}{7}$

7. Use indices in calculations.

e.g. if $y = x^2$ and $x = 3$, then $y = 3^2 = 9$

8. Give answers to an appropriate number of significant figures.

Give readings to the sensitivity of the instrument, e.g. 20 cm read from a metre rule should be 20.0 cm. Quote answers in calculations to as many significant figures as are given in the question, e.g. 23.56 is written as 24 to 2 significant figures.

9. Recognise direct and inverse proportion.

If y is directly proportional to x, y doubles when x doubles and a graph of y against x is a straight line through the origin.

If y is inversely proportional to x, y halves when x doubles and a graph of y against x is a curved line of negative decreasing gradient.

10. Interpret pie charts, bar graphs, and line graphs.

e.g. use pie charts to calculate fractions and percentages, read data from bar charts, and recognise trends from line graphs.

Mathematics for physics

11. Plot graphs, including a line of best fit.

Choose a suitable scale so that the points fill at least half of the graph paper.

A line of best fit can be a straight line or a curve. It does not need to pass through every point but there should be an even number of points above and below the line.

12. Find the gradient and intercept of a graph.

Draw a large right-angled triangle and use the scales on the axes to find the vertical height of the triangle (the 'y step') and the horizontal base of the triangle (the 'x step'). Divide the y step by the x step to find the gradient.

The intercept on the y axis is found using the scale when $x = 0$. The intercept on the x axis is found by reading from the scale when $y = 0$.

13. Recognise different shapes and terms used to describe them.

e.g. the diameter, radius, and circumference of a circle; squares, parallelograms, and rectangles; identify angles.

14. Find the area and volume of regular shapes.

shape	formula for area
rectangle	length × width
triangle	$\frac{1}{2}$ base × height
circle	π × (radius)^2

shape	formula for volume
rectangular block	length × width × height
cylinder	π × (radius)^2 × height

15. Recognise and use terms to identify different directions.

e.g. North, South, East, and West as points on the compass, clockwise and anti-clockwise for turning directions.

16. Calculate the sine of an angle or use inverse sine to calculate an angle.

e.g. $\sin(30°) = 0.5$, $\sin^{-1}(0.5) = 30°$

Mathematics for physics

17. Use appropriate symbols, units, and equations.

equation	symbols and SI units
$s = u \times t$	s = distance (m) u = speed (m/s) t = time (s)
$\Delta v = a \times t$	Δv = change in velocity (m/s) a = acceleration (m/s^2) t = time (s)
$F = m \times a$	F = force (N) m = mass (kg) a = acceleration (m/s^2)
$M = F \times d$	M = moment (Nm) F = force (N) d = perpendicular distance (m)
$KE = \frac{1}{2} \times m \times v^2$	KE = kinetic energy (J) m = mass (kg) v = velocity (m/s)
$PE = W \times h$	PE = gravitational potential energy (J) W = weight (N) h = height (m)
$F = p \times A$	F = force (N) p = pressure (Pa) A = area (m^2)
$E = P \times t$	E = energy (J) P = power (W) t = time (s)
$Q = I \times t$	Q = charge (C) I = current (A) t = time (s)
$V = I \times R$	V = potential difference (V) I = current (A) R = resistance (Ω)
$P = I \times V$	P = power (W) I = current (A) V = potential difference (V)
$m = \rho \times V$	m = mass (kg) ρ = density (kg/m^3) V = volume (m^3)

Mathematics for physics

equation	symbols and SI units
$v = f \times \lambda$	v = velocity (m/s) f = frequency (Hz) λ = wavelength (m)
$p = m \times v$	p = momentum (kg m/s) m = mass (kg) v = velocity (m/s)
$p = \rho \times g \times h$	p = pressure (Pa) ρ = density (kg/m³) h = height (m)
$E = m \times c \times \Delta T$	E = energy (J) m = mass (kg) c = specific heat capacity (J/(kg °C)) ΔT = change in temperature (°C)
$E = \Delta m \times L$	E = energy (J) Δm = change in mass (kg) L = latent heat (J/kg)
$\dfrac{\sin i}{\sin r} = n$	i = angle of incidence (°) r = angle of refraction (°)
$\sin c = \dfrac{1}{n}$	c = critical angle (°) n = refractive index (°)
$\dfrac{N_s}{N_p} = \dfrac{V_s}{V_p}$	N_s = number of turns on secondary coil N_p = number of turns on primary coil V_s = p.d. across the secondary coil (V) V_p = p.d. across the primary coil (V)

Exam-style questions

1. Figure 1 shows a distance–time graph for an object.

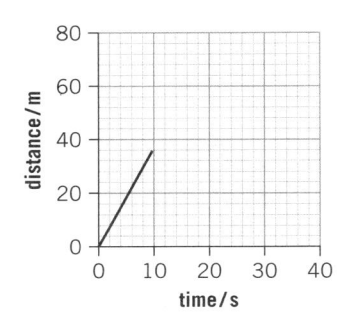

   Figure 1

   a. Describe the motion in the first 10 seconds, calculating any relevant quantity.

      ....

      .... [2]

   b. After 10 s the object decelerates for 20 s until it becomes stationary.

      Sketch on the figure the possible shape of the graph. [2]

   c. Figure 2 shows the corresponding speed–time graph for the object.

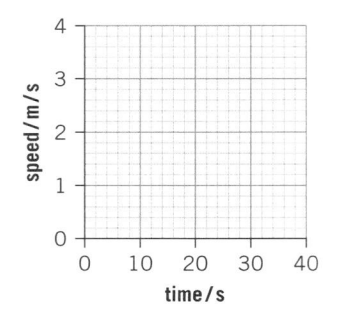

      Figure 2

      i. On Figure 2 draw the graph for the motion for the first 10 s as shown in Figure 1.

      ii. On Figure 2 draw the graph for the object decelerating at -0.1 m/s² for the next 30 s. [3]

   d. i. How can you tell if an object is stationary from its speed–time graph?

      ....

      .... [2]

      ii. Describe the difference between speed and velocity.

      ....

      .... [2]

   Total = 11

Exam-style questions

2. A simple pendulum swings with oscillations of constant time period.

   a. A student times single oscillations of the pendulum with a stopclock. He times an oscillation 8 times and obtains the following readings:

   1.2 s, 1.1 s, 1.2s, 1.2 s, 1.4 s, 1.3 s, 1.1 s, 1.2 s

   What is the best value obtainable from these readings for the time of one oscillation?

   Explain your reasoning.

   best value = ...

   explanation

   ....

   ....

   .... [1]

   b. Describe how, using the same stopclock, the student can find the time period of the oscillation more accurately.

   ....

   ....

   ....

   ....

   .... [4]

   Total = 5

3. A boy pulls his sledge up a hill of height 5 m and then slides down the other side. The mass of the boy and his sledge is 50 kg.

   a. What is the gain in potential energy as he climbs the hill?

   Potential energy = ... [2]

   b. i. What is the maximum kinetic energy he can have when he reaches the bottom of the hill?

   Kinetic energy = ... [1]

   ii. What is the maximum speed he can reach?

   Speed = ... [3]

   Total = 6

Exam-style questions

4. A student places a mass on a 'pressure toadstool', consisting of a circular wooden disk attached to a thin plastic block of square cross-section. She places the pressure toadstool into a tray of sand and measures how far it sinks into the sand. The arrangement is shown in Figure 3.

[Figure 3: diagram showing a 200 g mass on a circular wooden disk attached to a thin plastic block, placed in sand]

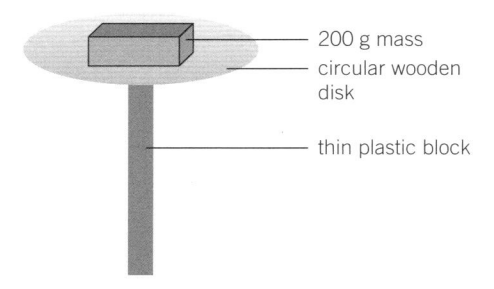

Figure 3

The cross-sectional area of the plastic block is 1.0×10^{-4} m². The mass of the pressure toadstool is 100 g.

a. Calculate the combined weight of the 200 g mass and the pressure toadstool.

weight = ... [2]

b. Calculate the pressure exerted on the sand.

pressure = ... [2]

c. When the 200 g mass is placed on the pressure toadstool, the student measures how far the thin plastic block sinks into the sand and finds the depth to be 22 mm.

Describe how the student could use a thin wooden rod, a pencil, and a 30 cm ruler to find an accurate value for the depth.

....

....

.... [2]

d. The student increases the mass on the pressure toadstool to 500 g and measures the depth to which it sinks into the sand. Predict the new depth and give a reason for your prediction.

depth = ...

reason ....

.... [3]

Total = 9

Exam-style questions

5. Figure 4 shows an object, labelled O, placed near a lens. The principal foci are labelled F.

   The two principal foci of the lens are F1 and F2.

   a. On Figure 4, draw the paths of two rays from the object so that they pass through the lens and continue beyond. Draw the location of the image and label it *I*. [3]

   b. Describe the image.

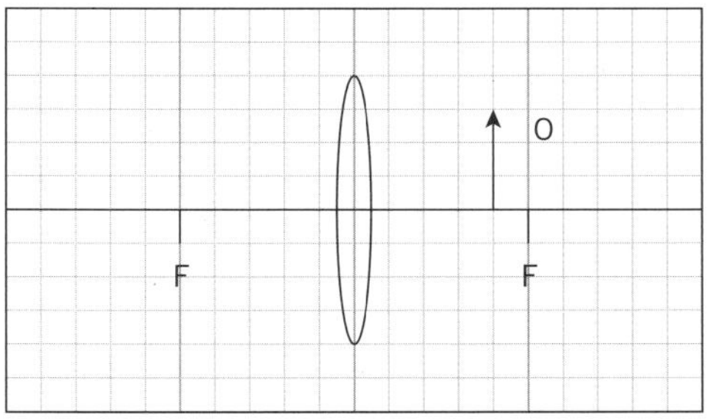

   Figure 4

   ....

   ....

   .... [3]

   c. Give a use for a lens in this arrangement.

   .... [1]

   Total = 7

6. a. Define *specific heat capacity*.

   ....

   .... [2]

   b. A vacuum flask holds 200 cm³ of warm water. A student measures the temperature of the water with a thermometer and finds it to be 35 °C. He adds cold water, with a temperature of 15 °C, a little at a time to the water in the vacuum flask until the temperature reaches 30 °C.
   Assume no thermal energy is lost or gained by the cup.
   specific heat capacity of water = 4.2 J / (g °C), density of water = 1 g / cm³

   i. Calculate the mass of warm water in the vacuum flask.

   mass of warm water = ... [2]

   ii. Calculate the thermal energy lost by the warm water in cooling from 35 °C to 30 °C.

   thermal energy lost = ... [2]

   iii. What is the thermal energy gained by the cold water in increasing its temperature from 15 °C to 30 °C?

   thermal energy gained = ... [1]

   iv. Calculate the mass of cold water added.

   mass of cold water added = ... [2]

   v. Hence calculate the volume of cold water added.

   volume of cold water added = ... [2]

   Total = 11

Exam-style questions

7. A water wave approaches a straight barrier with a gap in it. The width of the gap can be varied, as shown in Figure 5.

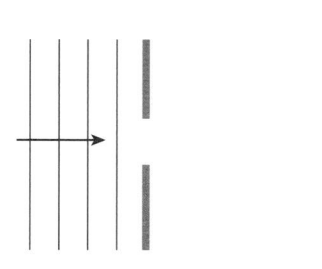

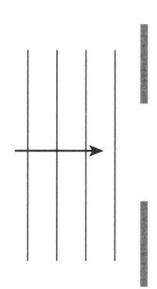

   Figure 5

   a. Draw the waves after they have passed through the gap on each diagram in Figure 5. [3]

   b. State the name given to this effect.

   .... [1]

   c. The water waves have a speed of 0.20 m/s and a wavelength of 10 cm. Calculate the frequency of the source producing the waves.

   frequency = ... [3]

   Total = 7

8. Four bulbs A, B, C, and D are connected so that A and B are in series with each other and C and D are in series with each other. A and B are in parallel with C and D respectively.

   a. Draw a circuit diagram of the arrangement, including a switch to control A and B and a switch to control all four bulbs and a 12 V power supply.

   [4]

   b. Each of the four bulbs is identical, with a power of 10 W.

      i. What is the potential difference across each bulb?

      Potential difference = ... [1]

      ii. Calculate the current through each bulb.

      Current = ... [2]

      iii. Calculate the charge passing through one of the bulbs in 3 minutes.

      Charge = ... [2]

   c. Calculate the energy transferred by one of the bulbs in 3 minutes.

   Energy = ... [2]

   Total = 11

Exam-style questions

9. The transformer in Figure 6 is designed to convert 230 V a.c. to 12 V a.c. The transformer is 100% efficient.

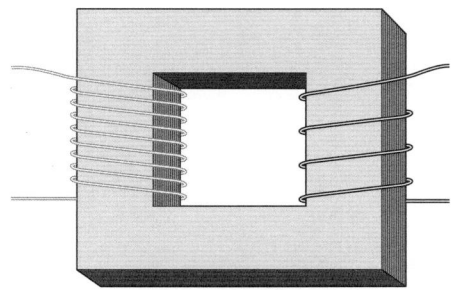

   Figure 6

   a. i. There are 500 turns on the primary coil. Calculate the number of turns on the secondary coil.

   number of turns = ... [2]

   ii. The input power is 50 W. State the output power.

   output power = ... [1]

   iii. Calculate the current in the secondary coil.

   current = ... [2]

   b. Describe how the transformer works.

   ....
   ....
   ....
   ....
   .... [3]

   Total = 8

10. Magnesium has many isotopes including the stable isotope magnesium-24 and the radioisotope magnesium-28.

    a. i. State one thing that is the same for atoms of the two isotopes.

    .... [1]

    ii. State one thing that is different for atoms of the two isotopes.

    .... [1]

    b. What is a radioisotope?

    ....
    .... [1]

153

c. Magnesium-28 decays into an isotope of aluminium by emitting a beta particle. Complete the decay equation.

$$^{28}_{12}\text{Mg} \longrightarrow \,^{\_}_{\_}\text{Al} + \,^{\_}_{\_}\beta$$

[4]

Total = 7

11. a. In the space below, draw the symbol for a NOR gate and label the inputs and the output.

[2]

b. Complete the truth table for the NOR gate.

Inputs		Output
0	0	
0	1	
1	0	
1	1	

[4]

c. Figure 7 shows a digital circuit made from a NOT gate and an AND gate.

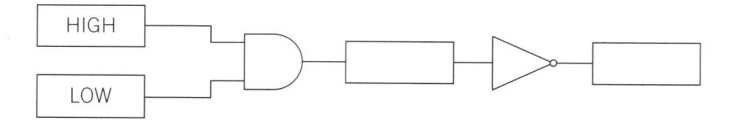

HIGH
LOW

Figure 7

  i. By looking at the logic gates, complete Figure 7 by writing the output (HIGH or LOW) in each of the boxes. [2]

  ii. State the effect on the output of changing both of the inputs.

  .... [1]

Total = 9

Project ideas

Investigate how the strength of a magnet varies with temperature

Using a bar magnet, design an experiment to vary the temperature of the magnet and measure its strength.

Investigate how the height to which a ball bounces depends on one or more of a number of factors

You may wish to study the temperature of the ball, surface onto which it falls, height of fall, or material from which the ball is made.

Investigate the factors that affect buoyancy

You could alter the size, shape, mass, or density of the object under test. Alternatively, you could investigate how the density of the liquid affects buoyancy, by changing the concentration of a salt solution for example.

Design a sleeping bag to keep someone warm

By considering your knowledge of the mechanisms of heat flow, design a sleeping bag that will reduce heat flow from a sleeping person. You can make and test your design by placing a hand warmer or a small jar filled with warm water in the model sleeping bag and measuring the rate at which the temperature falls.

Design an experiment to find out which colour surface is the best at absorbing heat

In your design, consider the effect of absorbed heat on a surface. Consider which variables you will need to control in your experiment and how you will control them.

Make a fruit battery

Using different fruit and vegetables and different combinations of metals (e.g. zinc, aluminium, copper) to use as electrodes, investigate how to make a battery and the factors that affect its voltage output.

Investigate friction

Investigate the frictional forces between different shoes and a surface. Consider how you will measure the frictional force and which factors you will investigate. Consider which variables you will need to control in your experiment and how you will control them.

Design a bridge

Research some bridge types and try to build your own from sections of rolled-up newspaper or drinking straws. Consider how you could test different bridge designs to find out which are strongest. Consider how you will make a comparison; for example, should they span the same distance or be made from the same mass of materials?

Answers

Unit 1.1
LL volume, measuring cylinder, metre rule, stop clock [4]
1. a. When the gun sounds, a timer starts automatically. [1] The timer stops when an athlete breaks a light beam at the finish line. [1]
   b. The race is short (less than 10 seconds in time) and the athletes finish very close together [1], so an automated timing system that can differentiate between them is needed. [1]
   c. The uncertainty due to the human reaction time would be too great to separate out the athletes. [1]
2. a. 50 ml [1]
   b. i. Place the metre rule against the block so that the zero lines up with a vertex. Measure the distance along that edge of the block. [1] Repeat for the other two dimensions. [1] Multiply the three values together. [1]
      ii. Vernier callipers [1]

Unit 1.2
LL If an object moves at a steady speed it covers the same distance every second. [1]
The average speed for a journey can be found by dividing the total distance by the time taken. [1]
The gradient of a distance–time graph is equal to the speed of the object. [1]
1. a. labelled axes [2]; points accurate to within ½ square [1]; line plotted which passes through all the points [1]
   b. 240 m [1]
   c. speed [1]
   d. i. steady speed [1]   ii. stationary [1]
2. a. distance increases with time [1] at a decreasing rate [1]; speed is decreasing to zero [1]
   b. distance increases with time [1] at an increasing rate [1]; speed is increasing [1]

Unit 1.3
LL The answer should describe a journey in terms of the distance covered and the time taken. The speed can then be presented. [3]
1. a. $s = v \times t$ [1]
   b. i. 2 m/s [1]   ii. 60 km/h [1]   iii. 20 s [1]
   c. i. 95 minutes
      ii. $10 \times 35 \times 60 = 21\,000$ m [1] $8 \times 55 \times 60 = 26\,400$ m [1]
      Total = 47 400 m [1]

Unit 1.4
LL acceleration, velocity, decelerate, distance, time, speed [6]
1. a. acceleration = change in velocity ÷ time [1]
   b. $a = [7-5] \div 0.5$ [1] $= 4$ [1]   m/s² [1]
   c. $t = [40-10] \div 3$ [1] $= 10$ [1]   s [1]
   d. change in velocity = 10 m/s [1]
      final velocity = 10 + 6 [1]   = 16 m/s [1]

Unit 1.5
LL If the velocity of a car is increasing this means it is accelerating. [1]
When a cyclist applies his brakes the bicycle will decelerate (slow down). [1]
The area under a velocity–time graph is equal to the distance travelled. [1]
1. a. i. distance travelled [1]
      ii. speed has size only [1]; velocity has both size and direction [1]
   b. i. section A $a = [10-0] \div 100$ [1] $= 0.1$ [1]   m/s² [1]
         section B $a = 0$ [1]   m/s² [1]
         section C $a = [0-10] \div 150$ [1] $= -0.067$ [1]   m/s² [1]
      ii. section A $s = ½ \times 100 \times 10$ [1] $= 500$ [1]   m [1]
         section B $s = 10 \times 150$ [1] $= 1500$ m [1]
         section C $s = ½ \times 10 \times 150$ [1] $= 750$ m [1]
      iii. 2750 m [1]

Unit 1.6
LL Across: 3. metre rule, 4. measuring cylinder, 5. metre, 7. Newton, 8. area, 10. volume, 11. stopclock, 14. force meter [8]
Down: 1. millimetre, 2. metres per second squared, 4. metres per second, 6. second, 12. balance, 13. joule [6]
1. a. Measure the equal widths with a rule. [1] Enter this into the data logger. Drop the card through the light gate. When the card interrupts the beam the timer starts. The timer stops when the beam is no longer interrupted. [1] The distance entered, divided by the time between starting and stopping of the timer, is the velocity for each portion of the card. [1] The data logger subtracts the two velocities and divides by the total time to calculate acceleration. [1]
   b. B is correct [1] Air resistance is negligible at such low speeds [1] and so the only force acting on the card is the weight, causing it to accelerate at 'g' m/s². [1]

Unit 2.1
LL
mass	the amount of matter in an object [1]
weight	force on an object in a gravitational field [1]
gravitational field strength	the force, due to gravity, on a mass of 1 kg [1]
kilogram	SI unit of mass [1]
newton	SI unit of weight [1]

1. a. Weight is the force on a mass in a gravitational field [weight = mass x gravitational field strength] [1]
   b. i. kg [1]   ii. N [1]
   c. 225.6, 542.4, 588.6, 226.2, 1416 [5]

Unit 2.2
LL The <u>density</u> of an object can be calculated if its <u>mass</u> and <u>volume</u> are known. Mass can be found using a <u>balance</u>. Volume can be found using a <u>measuring cylinder</u> if the object is <u>irregular</u>, or a <u>rule</u> if it is regular. [8]
1. a. kg/m³ [1] g/cm³ [1]
   b. balance [1]
   c. i. $180 \div 200$ [1] $= 0.9$ g/cm³ [1]
      ii. Yes [1] because the density of the oil is less than the density of water [1]
   d. mass = 25×13.5 [1] = 337.5 g [1]
2. a. height = 2.7 cm [1] radius = 1.4 cm [1]
      volume = $\pi r^2 h$ [1] = $\pi \times 1.4^2 \times 2.7$ [1] = 16.6 cm³ [1]
   b. mass = 16.6×7.9 [1] = 131.3 g [1]

Unit 2.3
LL Increasing the stretching force on a spring increases the extension. [1]
The extension of a spring is calculated by subtracting the original length from the new length. [1]
On a graph, the independent variable goes on the x-axis [1]
1. a. Diagram showing spring clamped and hanging vertically with mass suspended from the spring and rule clamped vertically next to it. [2] Without a mass suspended, take a reading of the position of the bottom of the spring using the rule. [1] Add a 100 g mass hanger to the spring. Take the new reading of the position of the bottom of the spring. [1] Repeat for further 100 g masses. [1] Subtract the original position from all subsequent readings to calculate the extension of the spring for each mass. [1]
   b. 0, 0.8, 1.6, 2.5, 3.5 [5] labelled axes [2] points plotted accurately to within ½ square [1] line of best fit drawn [1]
   c. Yes. [1] When the force doubles the extension roughly doubles [1] which indicates that extension is directly proportional to force. [1]

Unit 2.4
LL
friction	opposes motion when two surfaces rub together [1]
weight	the force due to gravity [1]
air resistance	a type of frictional force, which opposes motion in a fluid [1]
magnetic	the force between two poles [1]
electrostatic	the force between two charged particles [1]

1. a. Any three from: change direction, speed up, slow down, change shape [3]
   b. Any two from: friction, air resistance, drag [2]
2. A and B are both partially correct. [1] A body will remain stationary or move with constant speed in a straight line unless a force acts on it. [1] C has misunderstood that friction acts to oppose the motion so that with no driving force, the resultant force decelerates the object. [1]
3. i. 5 N [1]   ii. 3 N [1]   iii. 10 N [1]

Unit 2.5
LL resultant force, friction, motion, newton meter [4]
1. a. 0.75, 0.95, 0.6, 1.35 [1]
   b. 0.1 N [1]
   c. to increase the frictional force between the material and the smooth surface. [1]

Answers

  d. The frictional force depends on the speed/to ensure that the force due to the newton meter was equal to the force of friction. [1]
  e. Polythene bag. [1] Diagram should compare the size of the bumps in the surface of polythene, which are smaller than the bumps in the surface of other materials other materials. [1]
  f. Higher values [1] because only some of his pulling force balances friction. [1]

Unit 2.6
LL momentum — mass × velocity [1]
    impulse — change in momentum [1]
    velocity — speed in a certain direction [1]
1. a. kg m/s [1]
   b. i. 20 000 kg m/s [1]
      ii. 100 ÷ 70 [1] = 1.4 m/s [1]
      iii. 200 ÷ 4 [1] = 50 kg [1]
   c. i. 0.5 × [2.5–1.5] [1] = 0.75 kg m/s [1]
      ii. p = [1 × 0.6] + 1.6 [1] = 2.2 kg m/s [1]
          v = 2.2 ÷ 1 = 2.2 m/s [1]
      iii. Δp = 25 × 1.5 [1] = 37.5 kg m/s [1]
           v = 37.5 ÷ 30 = 1.25 m/s [1]

Unit 2.7
LL When a <u>bullet</u> is fired from a gun, it travels at great <u>speed</u>, which means that it has a large <u>momentum</u>. The Principle of <u>conservation</u> of Momentum predicts that the gun will have equal but <u>opposite</u> momentum to the bullet. Therefore the gun <u>recoils</u>. [6]
1. a. 0 kg m/s [1]
   b. 0.03 × 200 [1] = 6 [1] kg m/s [1]
   c. −6 kg m/s [1]
   d. 6 ÷ 0.70 [1] = 8.6 m/s [1]

Unit 2.8
LL force, impulse, momentum, time [4]
1. a. i. impulse F Δt = change in momentum M Δv [1]
      ii. kg m/s [1]
   b. i. (15 × 1000) − (25 × 1000) [1] = −10 000 kg m/s [1]
      ii. 325 − 250 [1] = 75 m/s [1]
   c. 82 000 × 0.03 [1] = 2460 kg m/s [1]
   d. i. (1200 × 11) ÷ 0.1 [2] = 132 000 N [1]
      ii. Stretchy fabric increases the time for the collision [1] which decreases the force which the passenger experiences [1] for the same change in momentum (since force = rate of change of momentum). [1]
2. a. 5 × 0.5 [1] = 2.5 N [1]
   b. 3000 ÷ 900 [1] = 3.3 m/s² [1]

Unit 3.1
LL Moment is equal to <u>mass</u> times <u>perpendicular</u> distance from the <u>pivot</u>, where force is measured in <u>newtons</u> and distance is measured in <u>metres</u>. A moment causes a <u>turning</u> effect on an object. [6]
1. a. 4.5 × 1.5 [1] = 6.75 Nm [1]
   b. 1.6 ÷ 0.8 [1] = 2.0 N [1]
2. a. i. 2 m [1] ii. F_1 = 12.5 N [1] F_2 = 7.5 N [1]

Unit 3.2
LL spring balance — a meter used to measure weight [1]
    equilibrium — when an object has no net moment or force acting on it [1]
    net moment — sum of all the clockwise and anticlockwise moments [1]
    pivot — the point about which an object can turn [1]
1. Multiply the value on the spring balance by 0.3 m (the distance of the spring balance from the pivot). [1] Multiply 9 N by 0.4 m (the distance of the weight from the pivot). [1] If the system is in equilibrium, clockwise moment = anticlockwise moment [1] and so the two calculated values should be equal. [1]
2. a. 7500 × 25 [1] = 187 500 [1] Nm [1]
   b. 187 500 Nm [1]
   c. 187 500 ÷ 50 000 [1] = 3.75 m [1]
   d. [50 000 × 5] ÷ 25 [1] = 10 000 N [1]

Unit 3.3
LL

E	C	N	A	T	S	I	D	F	C	H	N	N	M	N
K	Q	G	E	M	C	L	Z	L	O	Z	E	E	E	V
Q	D	U	K	C	D	I	O	X	T	R	S	T	T	Z
W	W	C	I	F	N	C	D	W	N	I	C	N	R	C
R	S	E	G	L	K	A	E	B	W	Z	D	E	E	N
O	J	W	M	W	I	N	L	K	G	O	N	M	U	W
D	B	U	J	E	T	B	C	A	F	N	V	O	T	Y
B	E	S	G	V	E	O	R	T	B	T	I	M	U	G
V	E	G	T	I	L	C	K	N	V	A	U	T	K	E
D	A	W	Y	C	N	X	H	U	U	L	Z	Z	D	G
M	S	P	I	V	O	T	I	V	R	M	W	U	S	L
J	I	T	N	B	X	X	M	J	M	V	Q	I	U	Q
N	N	V	R	U	X	R	L	U	B	U	V	W	R	U
A	L	F	A	U	S	E	E	H	M	H	R	O	O	Q
P	E	R	P	E	N	D	I	C	U	L	A	R	H	D

[12]

1. a. 15 × 30 [1] = 450 [1] N cm [1]
   b. 450 ÷ 3 [1] = 150 N [1]

Unit 3.4
LL moment, equilibrium, pivot, centre of mass [4]
1. [4]
2. Punch three holes in different places around the outside of the lamina. [1] Clamp a pin to a clamp stand and put the pin through one of the holes. [1] Hold a plumb line close to the pin so that the string hangs vertically downwards. [1] Mark the position of the string. [1] Repeat with each of the holes in turn. Where the lines cross is the position of the centre of mass. [1]

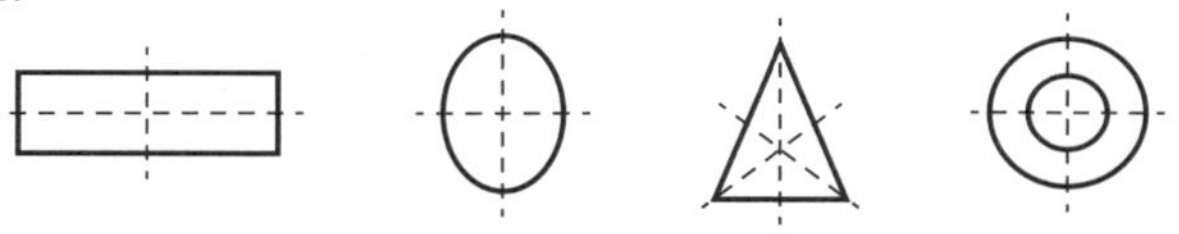

Unit 3.5
LL An object is <u>stable</u> if the centre of <u>mass</u> lies vertically above its base so that the line of action of the <u>weight</u> acts through the <u>base</u>. If the object is tilted so that the line of action of the weight no longer acts through the base, there will be a net <u>moment</u> on the object and it will <u>topple</u>. [6]
1. a. i. C [1]
      ii. The line of action of the weight acts outside the base [1] so there is a net moment. [1]
   b. i. It is more stable [1] because the centre of mass is lower [1] and therefore the weight is less likely to act outside the base. [1]
      ii. Having passengers on the top deck raises the centre of mass of the bus. [1] The weight will act outside the base for smaller angles of tilt [1], which causes a net moment. [1]
   c. The tractor has to be tilted through larger angles [1] for the line of action of the tractor's weight to act outside the base [1], in order to produce a net moment which would cause the tractor to topple. [1]

Unit 3.6
LL The <u>resultant</u> force is the <u>vector</u> sum of two or more forces. This takes into account both the <u>size</u> and <u>direction</u> of the forces. The resultant <u>force</u> can be determined by the use of an accurately drawn and scaled vector <u>diagram</u>. [6]
1. Resultant arrow drawn correctly, magnitude 19 N [2]
   Resultant arrow drawn correctly, magnitude 19 N [2]
2. Accurate scale diagram [2] answer = 4900 N [2]

157

Answers

Unit 4.1
LL

[Word search grid - 10 marks]

1.

chemical	stored in batteries and food
gravitational potential	gained as an object is moved away from the Earth
kinetic	energy an object has due to its movement
thermal	released when the temperature of a hot object decreases due to a decrease in its internal energy
strain	stored when an object changes shape
sound	produced by vibrating objects
light	emitted by very hot objects
nuclear	stored energy that can be released by fusion in the Sun
electrical	carried by moving charges in a circuit
Internal	total kinetic and potential energies of all the particles in an object

Unit 4.2
LL Anagrams: unjumble the key words
chemical, kinetic, sound, gravitational potential, electrical, thermal, nuclear [1]

1. a.

light bulb	electrical → light
electric iron	electrical → thermal
microphone	sound → electrical
waterfall	gravitational potential → kinetic
catapult	strain → kinetic
electric mixer	electrical → kinetic

[10]

   b. Energy cannot be created or destroyed, [1] rather it transforms from one form to another. [1]
2. a. 1 J, sound, heat [3]
   b. electrical, heat, 72 J, kinetic, sound [5]

Unit 4.3
LL

Beginnings:	Endings:
In a coal-fired power station, chemical energy	stored in coal is converted to electrical energy. [1]
Efficiency is the ratio of output energy	to input energy. [1]
Burning coal produces carbon dioxide which	contributes to the greenhouse effect. [1]
Burning coal produces sulphur dioxide which	causes acid rain. [1]
Some of the heat energy produced when coal is	burnt is dissipated into the surroundings. [1]
The efficiency of a coal-fired power station is	typically about 30%. [1]

1. a. coal, steam, kinetic, turbine, electrical, generator [6]
   b. nuclear energy from fission reactions, instead of chemical energy stored in coal [1]
   c. i. $2\,000\,000\,000 \div 0.3$ [2] $= 6\,700\,000\,000$ J [1]
      ii. heat [1]   iii. dissipated into the surroundings [1]
   d. carbon dioxide, sulfur dioxide [2]

Unit 4.4
LL In a nuclear power station, thermal energy is produced through nuclear fission. The thermal energy is used to heat water to produce steam, which turns the turbines. The turbines then turn the generator, which converts kinetic energy to electrical energy. [7]

1. work done in transporting fuel = change in kinetic energy [1] and kinetic energy $= \frac{1}{2} mv^2$ [1]. Carrying a higher mass of fuel requires more work to accelerate and decelerate the vehicle. [1]

Unit 4.5
LL hydroelectric, geothermal, fossil fuel, nuclear, unreliable, renewable [6]
1. kinetic, electrical, 5 J, thermal [4]
   Gravitational potential, 25 J, electrical, thermal [4]
2. $40 \div 100$ [1] $= 40\%$ [1]
   $350 \div 500$ [1] $= 70\%$ [1]
3. Wind turbines cause visual pollution. [1] Tidal barrages affect the habitats of sea birds [1]. Both require energy to build and hence carbon dioxide is produced. [1]

Unit 4.6
LL

renewable	an energy resource that will not run out [1]
non-renewable	an energy resource that will eventually run out [1]
reliable	an energy resource that produces electrical energy when required [1]
unreliable	an energy resource that is not always available [1]
fuel	a source of chemical energy that is burnt to release energy as heat [1]
energy resource	a means of producing electrical energy through energy transformations [1]
photovoltaic cell	a solar panel that transforms light energy to electrical energy [1]
fossil fuel	a store of chemical energy formed millions of years ago [1]

1. a. Renewable: wind, wave, tidal, geothermal, solar, hydroelectric, biofuel [7]
      Non-renewable: coal, oil, gas, nuclear [4]
   b. reliable: coal, oil, gas, nuclear, tidal, geothermal, biofuel [7]
      unreliable: wind, wave, solar, hydroelectric [4]
   c. require fuel: coal, oil, gas, biofuel, nuclear [5]
      do not require fuel: wind, wave, tidal, hydroelectric, solar, geothermal [6]

Unit 4.7
LL When the child slides down, his gravitational potential energy is transformed into kinetic energy. Some of his kinetic energy is dissipated into the surroundings as thermal energy due to work done against friction. [6]

1. a. $0.5 \times 50 \times 4^2$ [1] $= 400$ [1] J [1]
   b. $v^2 = 20 \div (0.5 \times 0.4)$ [1] $= 100$ [1] $v = 10$ m/s [1]
   c. $40 \times 10 \times 0.5$ [1] $= 200$ J [1]
   d. $105\,000 \div (70 \times 10)$ [1] $= 150$ m [1]

Answers

2. a. 1500 × 200 [1]   = 300 000 J [1]
    b. 40 000 ÷ 50 [2]   = 800 N [1]

Unit 4.8
LL power, efficiency, percentage, fraction. [4]
1. a. i. Watt (W) ii. $E = P \times t$ [1]
    b. i. efficiency = $\dfrac{\text{useful energy output}}{\text{energy input}}$ [1]
       ii. J ÷ J, hence no unit [1]
2. a. 9000 ÷ 30 [1] = 300 W [1]
    b. 3000 × 15 × 60 [2] = 2 700 000 J [1]
    c. 1 000 000 ÷ 25 [1] = 40 000 s [1]
3. a. 60 ÷ 200 [1] = 0.3 = 30% [1]
    b. 50 ÷ 0.8 [1] = 62.5 W [1]
    c. 4 ÷ 60 [2] = 0.067 = 6.7% [1]

Unit 5.1
LL When a force acts over an <u>area</u>, a <u>pressure</u> is exerted on that area. Force is measured in <u>newtons</u> and <u>area</u> is measured in m², so that <u>pressure</u> is measured in N/m² or <u>pascals</u>. [6]
1. a. i. N/cm² ii. N/m² or Pa [1]
    b. i. W = 0.8 × 10 = 8 N [1]
          8 ÷ 22 [1] = 0.36 N/cm² [1]
       ii. 20 × 0.4 [1] = 8 N [1]
    c. A = 10 × 10 = 100 cm² [1]
       15 ÷ 100 [1] = 0.15 N/cm² [1]

Unit 5.2
LL

Beginnings:	Endings:
The pressure in a liquid is	the same in all directions. [1]
As the depth of a liquid increases	the pressure increases. [1]
Pressure in a liquid is dependent on the	density of the liquid. [1]
The pressure deep in the ocean trenches is great	enough to crush a submarine. [1]

1. a. Using the pin, punch holes in the bag. [1] Fill the bag full of water and squeeze it. [1] Water will squirt out evenly in all directions. [1]
    b. As depth increases, pressure increases [1] at a steady rate. [1]
    c. 1000 × 10 × 2.25 [1] = 22 500 Pa [1]

Unit 5.3
LL manometer — an instrument used to measure the pressure due to a gas [1]
    pressure — the force per unit area [1]
    force — a push or a pull, measured in newtons [1]
    barometer — an instrument used to measure atmospheric pressure [1]
1. a. barometer
    b. The water reached boiling point [1] and changed from a liquid to a gas, dramatically increasing in volume. [1] The water vapour pushed the air out of the can. [1] When the can was rapidly cooled, the water vapour condensed back to a liquid [1] leaving a vacuum in the can. [1] The pressure from the air crushed the can. [1]
    c. Attach one arm of the manometer to the gas supply and turn on the gas. [1] The level in the two arms will no longer be equal. [1] The difference in pressure in the two arms is equal to the pressure of the gas supply. [1]

Unit 5.4
LL In a <u>solid</u> the particles are arranged in <u>rows</u> and are close together. When a solid <u>melts</u> to form a <u>liquid</u> the particles become randomly <u>arranged</u> but are still close together. When a liquid <u>boils</u> to form a gas, the particles become much further apart and move randomly in all <u>directions</u>. [7]
1. a. [diagrams showing solid, liquid, gas particle arrangements]

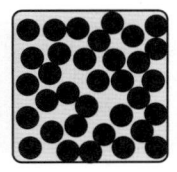

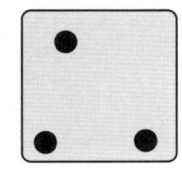

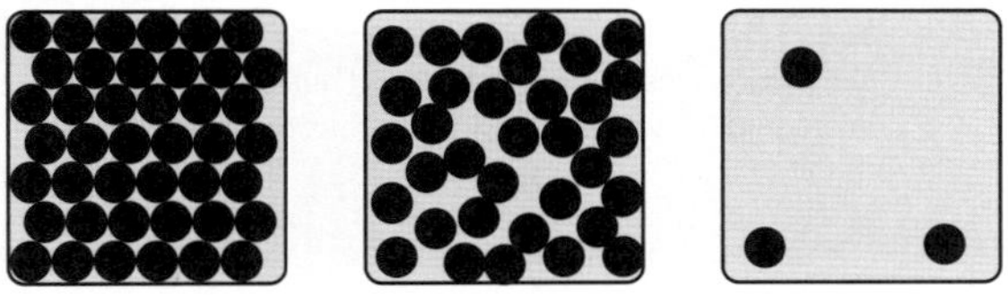

    b.
	solid	liquid	gas
	incompressible	incompressible	compressible
	cannot flow	can flow	can flow
	keeps its shape	takes the shape of the bottom of the container	fills the entire container
    [6]

    c. The particles are touching [1] and so cannot be pressed closer together. [1]
    d. Forces between the particles are weak enough [1] so that the particles can move past each other. [1]
    e. The smoke particles are in constant motion [1] in random directions [1] because they are being hit [1] by air particles travelling in constant random motion. [1]

Unit 5.5
LL Complete the sentences using the key words.
    In a solid the particles are <u>vibrating</u> in <u>fixed</u> positions. There are strong forces of <u>attraction</u> between particles and they are not <u>free</u> to move around. In a liquid the forces of attraction between particles are <u>weaker</u> than in a solid and so the particles are able to move around. In a gas the particles are very far <u>apart</u> and the forces of attraction between them are <u>negligible</u>. [8]
1. a. Particles in ice are close together in fixed positions. [1] As the ice melts to become water, particles are still close together but randomly arranged. [1] As water evaporates, particles leave the liquid surface and are far apart and randomly arranged. [1]
    b. Particles in ice vibrate in fixed positions. [1] As the ice melts to become water, particles are able to move past each other. [1] Once evaporation has occurred, particles are moving randomly in all directions. [1]
2. During collisions between gas particles and the walls of the container [1], the momentum of the gas particles changes. [1] Hence particles exert a force on the container walls. [1] Since pressure = force ÷ area, the particles exert a pressure on the walls of the container. [1]

Unit 5.6
LL velocity — distance travelled per second in certain direction [1]
    heat — a type of energy that raises the temperature of an object [1]
    pressure — the force per unit area [1]
    volume — the amount of space a substance takes up [1]
    particle — an atom or a molecule [1]
    smoke cell — a small glass box containing air and other particles [1]
1. a. As the temperature of the gas increases the average kinetic energy of the particles increases [1] and so the particles move faster. [1] This increases the force with which they hit the walls of the container [1] and the frequency of the collisions with the walls of the container [1], which increases the pressure. [1]
    b. i. gases [1]
       ii. In gases the particles are further apart than in solids and liquids. [1] The increase in kinetic energy causes the particles to move faster and hence move further apart. [1]

Unit 5.7
1. a. Some particles from the surface of the liquid [1] are evaporating [1] to form a vapour or gas. [1]
    b. The particles with the greatest kinetic energy [1] are moving fast enough [1] to escape the forces of attraction between particles in the liquid. [1]
    c. The vapour cools [1], the particles slow down [1] and move closer together forming a liquid. [1]
2. a. Wrap cotton wool around the bulb of each thermometer. [1] Soak one in acetone, one in water and leave one dry. [1] Start the stop clock simultaneously. [1] Measure the temperature on each thermometer at 30 s intervals. [1] Plot temperature against time and compare the cooling curves. [1]

159

Answers

Unit 5.8
LL
apparatus	equipment for an experiment [1]
accurate	close to the true value [1]
pressure gauge	an instrument used to measure pressure [1]
relationship	how one variable affects another [1]

1. a. Increase the pressure using the pump [1] and measure the value of pressure on the pressure gauge. [1] Read the volume with eye level to the meniscus. [1] Repeat each reading and average after discarding any anomalous results. [1]
   b. Axes accurately drawn and labelled. [2] All points plotted accurately. [1] Accurate best fit straight line, with negative gradient. [1]
   c. Student takes two values of pressure, one which is half the size of the other (e.g. 200 kPa and 100 kPa). Reads off the corresponding values of volume. [2] Shows that the higher pressure has half the volume of the lower pressure. [1]

Unit 6.1
LL temperature, increases, expands, buckle [4]
1. When the temperature increases the metal track expands. [1] This closes the gap. [1] Otherwise the sections would push against each other and buckle. [1]

Unit 6.2
LL thermistor, thermocouple, mercury-in-glass, Celsius [4]
1. a. alcohol-in-glass, thermistor, thermocouple, mercury-in-glass (any 3) [3]
   b. 0°C and 100°C [2]
   c. they expand on heating [1]
   d. Advantage: gives greater precision/ do not need to take readings manually/ can be left in remote locations. [1]
   Disadvantage: more expensive/more difficult to set up. [1]
2. a.

	absolute zero	melting ice	boiling water
Kelvin scale	0 K	273 K	373 K
Celsius scale	−273 °C	0 °C	100 °C

[4]
   b. Theoretical temperature below which [1] all particles cease to vibrate. [1]
   c. Particles gain kinetic energy. [1] Since kinetic energy = $\frac{1}{2}mv^2$ [1], particle velocity increases. [1]

Unit 6.3
LL
capillary tube	very thin glass tube [1]
test tube	a thin glass tube closed at one end [1]
boiling	turning from a liquid to a gas at constant temperature [1]
bung	a rubber stopper used to close a test tube [1]
calibrate	mark with a scale [1]

1. a. At 0 °C the level of oil will fall below the original level. [1] At 100 °C the level of oil will rise above the original level. [1]
   b. Mark the level when the test tube is in ice water at 0 °C. [1] Mark the level when the test tube is in boiling water at 100 °C. [1] Divide up the space between these marks into equally spaced lines, marked with temperatures. [1] Place the test tube in the warm water and use the level against the marks to read the temperature. [1]

Unit 6.4
LL specific heat capacity, thermometer, electrical heater [3]
1. a. energy required to raise temperature of 1 kg of material by 1°C [1]
   b. J/kg °C [1]
   c. i. 2400 ÷ (0.4 × 15)   [1]   = 400 J/kg °C   [1]
      ii. 0.5 × 10 × 4200   [2]   = 21 000 J   [1]
      iii. 2000 ÷ (0.2 × 350)   [2]   = 29°C   [1]
2. a. energy = 50 × 16 × 60 = 48 000 J   [1]
      c = 48 000 ÷ (0.5 × 22)   [2]   = 4400 J   [1]
   b. energy = 50 × 16 × 60 = 48 000 J   [1]
      c = 48 000 ÷ (0.5 × 22)   [2]   = 4400 J   [1]

Unit 6.5
LL Match the key words with their meanings.
beaker	a glass container used to contain liquids [1]
Bunsen burner	a source of heat, which uses natural gas as a fuel [1]
constant	not changing [1]
temperature	the degree of 'hotness', related to an object's internal energy [1]

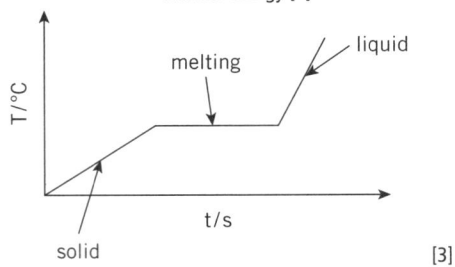

b. i. see labels on the graph [3]
   ii. As energy is supplied at a constant rate [1] temperature increases at a constant rate. [1]

Unit 6.6
LL When a substance is heated, the <u>internal</u> energy increases and its <u>temperature</u> rises. <u>The specific heat capacity</u> is the energy required to raise the temperature of 1 kilogram of the material by 1 °C. The <u>mass</u> of material can be found using a balance and the temperature rise measured with a <u>thermometer</u>. [5]
1. a. 0.45 × 2 300 000   [1]   = 1 040 000 J   [1]
   b. 65 400 ÷ 0.6   [1]   = 109 000 J/kg   [1]

Unit 6.7
LL
melt	change state from solid to liquid [1]
solidify	change state from liquid to solid [1]
conductor	passes on heat energy by particle vibrations [1]
insulator	does not easily pass on heat energy by particle vibrations [1]
free electron	a negatively charged particle that is not bound in an atom [1]
ion	an atom that has lost one or more electrons [1]

1. a. They fall off [1] with the ones nearest the flame falling first. [1]
   b. Wood is a poor conductor [1] so the heat cannot conduct away and the wooden bar becomes hot. [1]
   c. An atom that has gained or lost an electron. [1]
   d. The electrons gain kinetic energy and move around faster. [1] They collide with the metal ions, passing on the energy and causing the ions to vibrate faster. [1]

Unit 6.8
LL convection, density, separation, Bunsen burner [4]
1. a. [2]
   b. As the water is heated the particle separation increases [1], decreasing the water's density and causing the water level to rise. [1]
   c. The particles lose kinetic energy and move closer together [1], increasing the water's density and causing the water level to fall. [1]

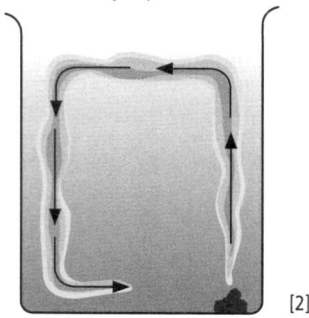

Answers

Unit 6.9

LL Match the key words with their meanings
- infrared radiation — heat that travels in electromagnetic waves [1]
- emission — giving out [1]
- absorption — taking in [1]
- reflection — bouncing back [1]

1. a. matt black, shiny black, white, silver [2]
   b. 13, 4
   c. black test tube [1] experiences the greatest rise in temperature [1]

Unit 6.10

LL [wordsearch grid] [10]

1. vacuum – reduces conduction and convection [1] because it does not contain particles. [1]
   steel walls with silvery surfaces – reduces radiation [1] because silver is a poor absorber/ good reflector of infrared radiation. [1]
   stopper – reduces evaporation [1] because the water vapour cannot escape from the top of the flask. [1] Also reduces convection [1] because warm air cannot rise out of the flask. [1]

Unit 7.1

LL [wordsearch grid] [7]

1. Longitudinal [1]
2. a. i. The number of waves passing a point per second/ number of oscillations per second. [1]
      ii. Hertz (Hz) [1]
   b. speed = frequency × wavelength
   c. 2.3 × 100 [1]   = 230 m/s [1]

Unit 7.2

LL wavelength, amplitude, vibration, oscilloscope, transverse, longitudinal, transfer, evacuated [8]

1. a. Longitudinal: The coils oscillate [1] parallel to the direction of wave travel. [1]
      Transverse: The coils oscillate [1] at right angles to the direction of wave travel. [1]
   b. i. wavelength [1]   ii. amplitude [1]

Unit 7.3

LL As a water wave travels from <u>deep</u> to shallower water, it <u>slows</u> down and its wavelength <u>decreases</u>. The <u>wavefronts</u> get closer together. The frequency of the wave is <u>unchanged.</u> [5]

1. a. i. ii. [diagram] [4]
      iii. As the wave enters the shallower water it slows down [1] and therefore its wavelength decreases [1] and the direction of travel bends towards the normal. [1]
   b. 90° [1]
   c. decreases [1]
2. a. i. 20–20 000 Hz [1]
      ii. ultrasound [1]

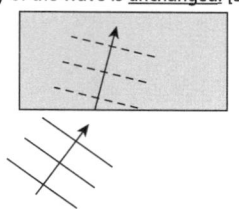

Unit 7.4

LL Across: 1 transverse, 3 ripple tank, 6 compression, 8 metre, 10 longitudinal, 13 wavefronts, 15 frequency [7]
Down: 2 rarefaction, 4 hertz, 5 wavelength, 7 sound, 9 water, 11 amplitude, 12 speed, 14 energy. [8]

1. [diagrams] [1] [1] [1]
diffraction [1]

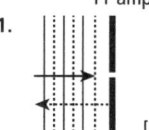

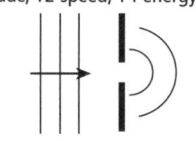

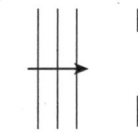

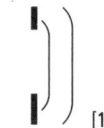

Unit 8.1

LL
- plane mirror — a flat reflective surface [1]
- normal — a constructed line at right angles to a surface [1]
- angle of incidence — angle between the incident ray and the normal [1]
- angle of reflection — angle between the reflected ray and the normal [1]
- real — an image that can be focused on a screen [1]
- virtual — an image that cannot be focused on a screenray [1]
- ray — a narrow beam of light [1]

1. a. To avoid looking into the laser beam, which could damage their eyes. [1]
   b. Laser light travels in a straight line. [1] The powder reflects the light into their eyes. [1]
2. a. virtual, upright, same size as the object [3]
   b. A virtual image is formed behind the mirror and cannot be focused on a screen. A real image can be focused on a screen. [1]

Unit 8.2

LL When light travels from air to glass, it <u>slows</u> down and bends towards the <u>normal</u>. This is due to the change in <u>density</u> of the material. When light travels from glass to air, it <u>speeds</u> up and bends <u>away</u> from the normal. This phenomenon is called <u>refraction</u>. [6]

1. a. Bending of light rays due to a change in their speed. [1]
   b. decreases [1]

161

c.

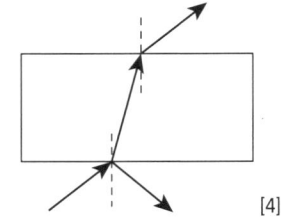

[Diagram: light ray entering and exiting a rectangular block showing refraction] [4]

2.

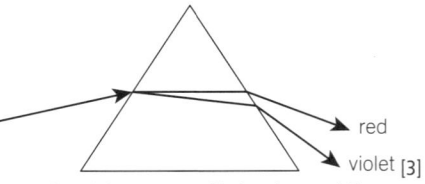

[Diagram: prism splitting light into red and violet] [3]

a. White light consists of light of many different wavelengths. [1] Different wavelengths travel at different speeds and are refracted by different amounts. [1] Violet light diffracts the most because it has the shortest wavelength. [1]

b. diffraction [1]

Unit 8.3
LL refraction, transparent, refractive index, speed of light [4]

1. a. $\frac{\sin i}{\sin r} = n$ [1]
   b. i. $\sin 45 \div \sin 30$ [1] $= 1.4$
      ii. $\sin r = \sin 40 \div 1.4$ [1]
          $r = \sin^{-1}(\sin 40 \div 1.4)$ [1] $= 27°$ [1]
      iii. $\sin i = 1.6 \times \sin 25$ [1]
           $i = \sin^{-1}(1.6 \times \sin 25)$ [1] $= 43°$ [1]
   c. $i = 64°$ [1], $r = 31°$ [1],
      $\sin 64° \div \sin 31°$ [3] $= 1.7$ [1]
   d. i. $\sin c = 1 \div n$ [1]
      ii. $1 \div \sin 55$ [1] $= 1.2$ [1]
      iii. $\sin c = 1 \div 1.5$ [1]
           $c = \sin^{-1}(1 \div 1.5)$ [1] $= 42°$ [1]

Unit 8.4
LL
refracted ray	beam of light that has changed direction on entering a different medium [1]
critical angle	the incident angle beyond which light is totally internally reflected [1]
total internal reflection	when all light rays are reflected inside a more dense medium [1]
optical fibre	very thin glass rod down which light travels by total internal reflection [1]
boundary	where one medium ends and another begins [1]

1. a. Total internal reflection. [1] All light would be reflected at the boundary and none would emerge from the block. [1]
   b. i. [Diagram of optical fibre with cladding and core, showing TIR] [5]
      ii. A bundle of fibres illuminates the inside of the body. [1] Light returns to the observer by TIR in another bundle of fibres. [1] An image is formed from this light. [1]
      iii. Each fibre provides another 'pixel' of the image. [1] Having many fine fibres allows a brighter image [1] and gives a sharper image. [1]

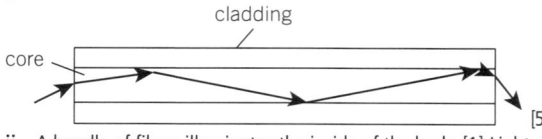

Unit 8.5
LL When light rays which were originally parallel and close to the principal axis pass through a convex converging lens, they converge to a point called the principal focus. The distance between the centre of the lens and the principal focus is called the focal length. [6]

1. a.

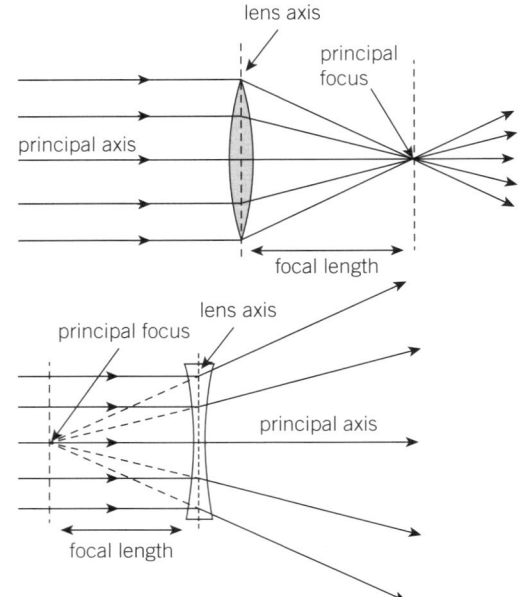

[Diagram: convex lens showing parallel rays converging at principal focus, with lens axis, principal axis, focal length labelled]

[Diagram: concave lens showing parallel rays diverging, with principal focus and focal length labelled]

b. i. ii.

[Ray diagram: convex lens with object beyond 2F, image formed between F and 2F on other side]

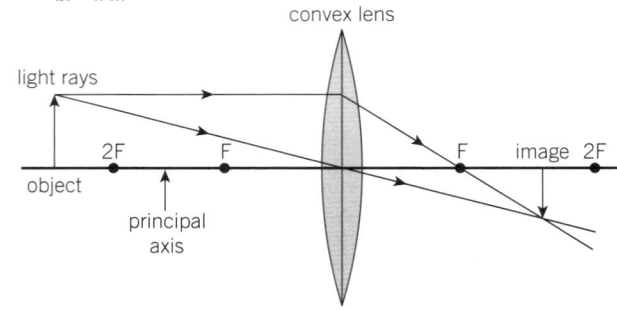

real, inverted, diminished [2]

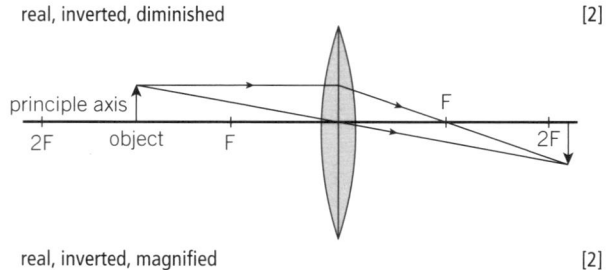

[Ray diagram: convex lens with object between F and 2F, image formed beyond 2F]

real, inverted, magnified [2]

c. i. camera [1]   ii. projector [1]

Unit 8.6
LL diminished, magnified, real, upright, virtual, inverted [6]

1. a. i. [2] ii. [1]. iii. [4]

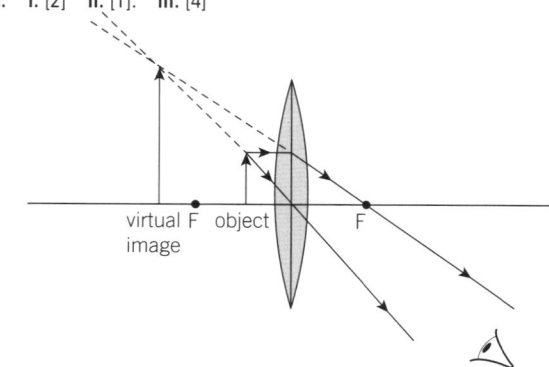

[Ray diagram: convex lens with object between lens and F, virtual image on same side as object]

b. virtual, upright, magnified [3]

162

Answers

Unit 8.7
LL microwave, ultraviolet, infrared, radio [4]
1. a. transverse waves [1] travel at 3×10^8 m/s in a vacuum [1] can be reflected/refracted/diffracted [1]
   c. i. gamma [1]   ii. radio [1]
      iii. infrared, radio, microwaves, visible (any three) [3]
      iv. UV, X-ray, gamma [3]
   d. i. imaging bones [1]   ii. gamma [1]
      iii. UV [1]   iv. microwave [1]
   e. 300 000 000 ÷ 100 [1] = 3 000 000 Hz [1]

Unit 8.8
LL

(word search grid) [10]

1. a. X-rays incident on a patient are absorbed by bone. [1] They are not absorbed by soft tissue. [1] A shadow picture of the bones is formed on the film. [1]
   b. Sunbeds produce UV, which is ionising radiation. [1] UV is absorbed in the skin and ionises atoms in cells. [1] This can lead to cell damage/death/cancer developing. [1]
   c. Humans emit more infrared radiation than objects [1] because they are at a higher temperature. [1] At night, when there is little visible light to track criminals, police can track a criminal's infrared radiation. [1]

Unit 9.1
LL
- tuning fork — a vibrating metal instrument that produces a sound of constant frequency [1]
- vibration — back and forth motion [1]
- evacuated — not containing any particles [1]

1. a. The tuning fork prongs vibrate [1] at a constant frequency [1]. He hears a note of that frequency. [1]
   b. No sound would be heard [1] because sound cannot travel in a vacuum. [1]
2. a. 0.1 s [1]
   b. 1500 × 0.1 [1] = 150 m [1]

Unit 9.2
LL compression, rarefaction, oscilloscope, amplitude [4]
1. a. i. (diagram showing direction of travel, wavelength, compression, rarefaction) [3]
      ii. high pressure at a compression [1] to low pressure at a rarefaction [1]
      iii. 330 ÷ 2000 [1] = 0.165 m [1]
   b. i. Diagram must show new wave with double the wavelength [1] and $\frac{1}{3}$ the amplitude. [1]
      ii. lower pitch [1] and quieter [1]

Unit 9.3
LL When the hammer strikes the metal block a sound <u>wave</u> travels to <u>microphone A</u> and <u>starts</u> the timer. When the sound reaches <u>microphone B</u>, the timer <u>stops</u>. The student can calculate the <u>speed</u> by dividing the distance between the two microphones by the <u>time</u> on the timer. [7]
1. a. Sound reaching A switches on the timer. [1] When the sound has travelled 1 m it reaches B and switches off the timer. [1] The length measured by the ruler divided by the time is equal to the speed of sound in air. [1]
      i. s [1]
      ii. The ruler can be read to the nearest mm. [1]
      iii. to increase accuracy of measurements by taking an average [1]
      iv. 0.0030 s [1]
      v. 329 [1]   m/s [1]
   b. 330 × 1 [2] = 330 m [1]

Unit 9.4
LL
- noise — sound waves that vary randomly in frequency and amplitude [1]
- vocal cord — the vibrating membranes that produce sound in humans [1]
- percussion — an instrument that produces a sound when struck [1]
- amplitude — the maximum displacement of a particle from the equilibrium position [1]
- pitch — the frequency of the sound — that is, how high the note sounds [1]
- oscilloscope — used to display the waveforms of sound waves [1]
- signal generator — used to produce waves of varying amplitude and frequency [1]

1. a. As the amplitude of the waves produced by the signal generator increases [1], the height of the waveform on the screen increases [1] and she hears a louder sound [1].
   b. The higher the frequency of the waves produced by the signal generator [1], the higher the pitch of the sound she hears [1] and the more waves appear on the screen. [1] The waves on the screen are closer together. [1]
2. a. voice – vibrating vocal cords [1]
      string – transverse vibrations of the string [1]
      tube – longitudinal vibrations of the air in the tube [1]
   b. The voice and the string would have more overtones than the tube. [1]

Unit 10.1
LL magnet, north pole, south pole, ferromagnetic [4]
1. a. i. attracts [1]   ii. repels [1]
   b. iron [1]
2. A – unmagnetised iron   B – non-magnetic material
   C – magnetised iron [3]

Unit 10.2
LL
- attract — pull towards with a force [1]
- repel — push away with a force [1]
- magnet — a magnetised piece of iron [1]
- ferromagnetic — a particular type of magnetism most commonly found in iron [1]
- domain — a tiny magnet found inside a magnetic material [1]
- unmagnetised — a ferromagnetic material in which the domains are not lined up [1]
- magnetic field — the area around a magnet where another magnet experiences a force [1]

1. a. Stroking with a magnet/ place in an electromagnet [1]
   b. Hitting/place inside electromagnet carrying AC current [1]

Answers

2. a. **magnetic** field [1]
    b.

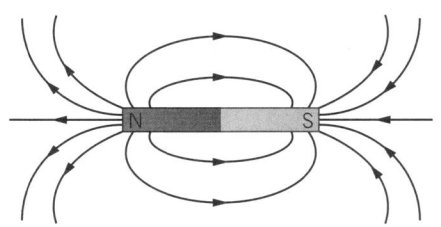

    [3]

    c. The direction a north pole would move if placed in the field. [1]
    d. This is where the magnetic field is stronger. [1]
3. a. [3]   b. [3]

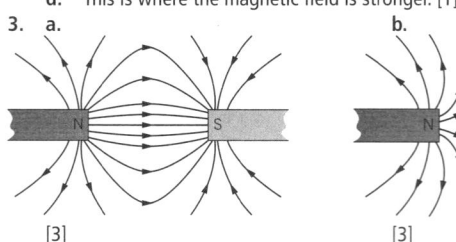

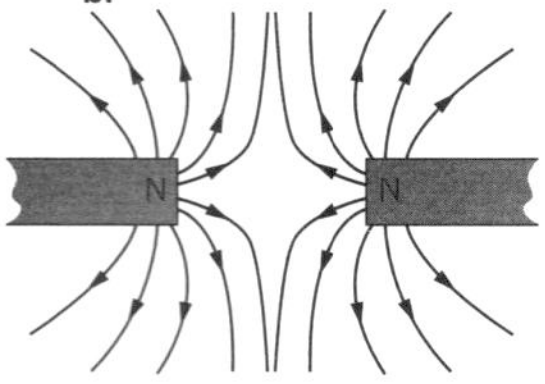

Unit 10.3
LL

[wordsearch grid] [9]

1. a. Vary the resistance of the variable resistor [1] to change the current flowing in the circuit. [1] For increasing values of current, measure how many connected paperclips the electromagnet can hold. [1] The greater the number of paper clips, the greater the magnetic field. [1]
    b. label axes [2] plot points accurately [1] line of best fit [1]
    c. number of paper clips doubles [1]
    d. 13 or 14 [1]

Unit 11.1
LL
charged	containing an unbalanced number of positive and negative charges [1]
coulombmeter	used to measure charge [1]
electrostatic	the type of force between charged particles [1]
insulator	contains charged particles that are not free to move [1]
coulomb	the unit of charge [1]

1. a. i. positive on cotton, negative on polythene [1] negative on silk, positive on acetate [1]
       ii. electrons [1]
       iii. friction [1]
       iv. coulomb [1]
       v. from cotton to polythene [1] from acetate to silk [1]
    b. i. attracts [1]   ii. rods would repel each other [1]

Unit 11.2
LL

[wordsearch grid] [9]

1. a. The direction in which a positive charge would move in the electric field. [1]
    b.

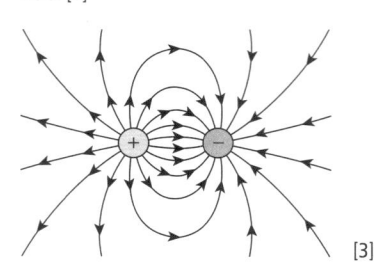

    [3]

Unit 11.3
LL

Beginnings:	Endings:
A positively charged object	has lost electrons. [1]
A negatively charged object	has gained electrons. [1]
Charge is transferred between objects by	the force of friction. [1]
The arrows on electric field lines show the	direction a positive charge would move. [1]

1.

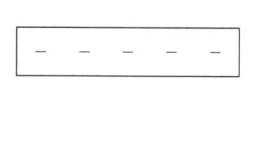

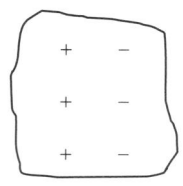

    [2]

    The electrons in the paper are repelled from the rod, leaving a positive charge near the rod. [1]
2. a. The force of friction [1] between the air and the aeroplane [1] causes transfer of electrons between the air and the aeroplane. [1]
    b. The aeroplane is earthed by the metal wire [1] so charge can flow off the plane to ground. This prevents sparks caused by discharging to ground through the air [1] which could set the fuel on fire. [1]

Unit 11.4
LL current, ammeter, ampere, series [4]
1. a. i. charge passing a point each second [1]
       ii. Amp (A) [1]
       iii. ammeter [1] —(A)—
    b. flow of free electrons [1]
    c. i. 2×10 [1] $= 20$ C [1]
       ii. 3×0.01 [1] $= 0.03$ C [1]

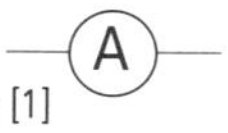

164

Answers

    iii. 30 × 60 × 0.05 [1] = 90 C [1]
    iv. 120 ÷ (1 × 60) [2] = 2 A [1]
    v. 0.018 ÷ 6 [2] = 0.003 A [1]
    vi. 100 ÷ 2 [1] = 50 s [1]
    vii. 0.01 ÷ 0.005 [2] = 2 s [1]

Unit 12.1
LL Batteries range in size from the tiny, low <u>voltage</u> batteries found in calculators to the large batteries used to start a car. A battery consists of two or more <u>cells</u> connected together. Batteries transform <u>chemical</u> energy to <u>electrical</u> energy. One side of the battery is <u>positively</u> charged and the other side is <u>negatively</u> charged. In a <u>complete</u> circuit, the opposite charges of the battery cause charges in the circuit to move and hence a <u>current</u> flows. The electromotive force of a cell is measured in volts and is the maximum <u>potential difference</u> that a cell can supply. The higher the electromotive force of a cell, the greater the <u>energy</u> given to each charge as it passes through the cell. [10]

1. a. independent variable – number of cells [1] dependent variable – potential difference across the cells [1] controlled variable – e.m.f. of each cell [1]. Potential difference measured with a voltmeter. [1] Increase the number of cells connected in series by one each time, and measure the p.d. across all cells. [1]
   b. [graph: p.d. across all cells/V vs number of cells, straight line through origin] [4]

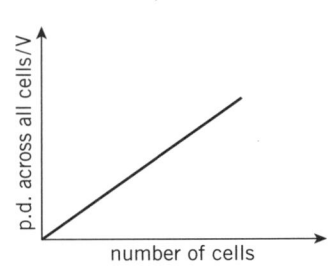

Unit 12.2
LL voltage, potential difference, voltmeter, battery [4]
1. a. i. work done per coulomb of charge [1]
    ii. volt [1]
    iii. voltmeter —(V)— [1]
  b. 4.5 V [1]
  c. 2V across each bulb [1] 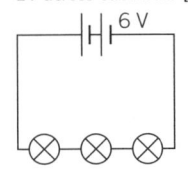
    [circuit diagram with 6V battery and three bulbs] [1]

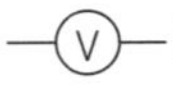

Unit 12.3
LL resistor, resistance, current, proportional [4]
1. a. Vary the p.d. across the fixed resistor [1] by moving the slider on the variable resistor. [1] Measure the p.d. across the fixed resistor using the voltmeter. [1] Measure the corresponding current across the fixed resistor using the ammeter. [1]
  b. no unit given for potential difference [1] no label for x-axis [1] axes wrong way round (x-axis should show p.d., and y-axis should show current) [1] On both axes, 1 should be written as 1.0 [1]
  c. i. 0.45 V [1]   ii. 0.9 A [1]
  d. yes [1]; as the p.d. doubles, the current also doubles. [1]
  e. The relationship holds for voltages outside this range, provided that the resistor is at constant temperature. [1] At much higher voltages, the resistor may heat up and then the relationship would no longer hold. [1]
  f. The free electrons which form the current [1] collide with [1] vibrating metal ions in the resistor, which opposes their motion. [1]

Unit 12.4
LL
resistance	potential difference required to make one ampere of current flow [1]
current	the charge passing a point per second in a circuit [1]
potential difference	the work done on a coulomb of charge between two points in a circuit [1]
ammeter	used to measure current in a circuit [1]
voltmeter	used to measure potential difference in a circuit [1]
cell	a source of chemical energy that causes current to flow around a circuit [1]

1. a. [circuit diagram with ammeter A, constantan wire, and voltmeter V] [3]
  b. metre rule [1]
  c. Ω [1] 1.0, 2.0, 2.9, 4.0, 4.9, 6.1, 7.1 [2]
  d. Yes [1] as the length of wire increases, the values of resistance also increase. [1] Resistance increases at a steady rate. [1]
2. a. V = I × R [1]
  b. 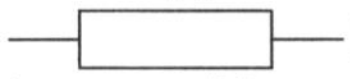 —[resistor symbol]— [1]
  c. i. 3 × 6 [1] = 18 V [1]   ii. 12 ÷ 0.5 [1] = 24 Ω [1]

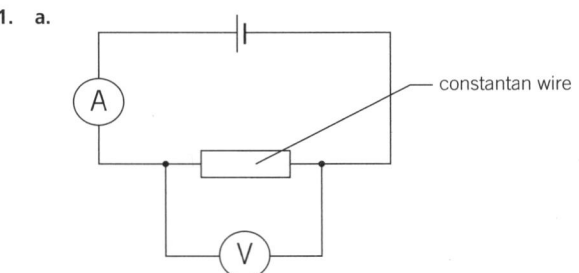

Unit 12.5
LL Work is done on charges as they pass through a cell and they gain <u>energy</u>. As the charges pass through a bulb in the <u>circuit</u>, they pass their <u>energy</u> to the metal <u>ions</u> in the bulb. The ions <u>vibrate</u> more and the filament becomes hotter. The bulb gives out <u>light</u> and heat energy. The work done per second on the bulb is the <u>power</u>. [7]
1. a. i. P = I × V [1]   ii. watt (W) [1]
  b. i. 40 ÷ 12 [1] = 3.3 A [1]
    ii. 2 × 10 [1] = 20 W [1]
    iii. 60 ÷ 0.5 [1] = 120 V [1]

Unit 13.1
LL thermistor, variable resistor, light dependent resistor, diode [4]
1. a. 1st column: thermistor, diode, LDR
    2nd column: cell, variable resistor, resistor [6]
  b. [circuit diagram with thermistor, ammeter A, voltmeter V, and beaker of water with thermometer on bunsen burner] [4]
  c. Heat the water with the Bunsen burner. [1] Measure the temperature of the water with the thermometer (stir the water). [1] Take readings of current from the ammeter and voltage from the voltmeter [1] at 10°C intervals. Divide voltage by current to find resistance. [1]

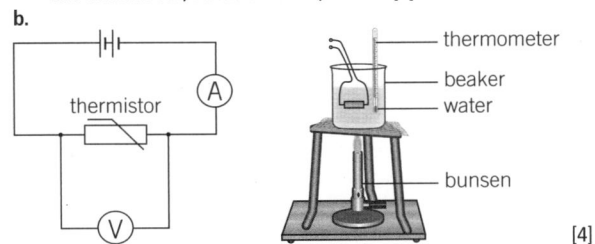

Unit 13.2
LL In a circuit, a <u>voltmeter</u> is placed in <u>parallel</u> with the component, to measure the potential difference, and an <u>ammeter</u> is placed in series with the component to measure the current. The potential difference across the component can be varied using a <u>variable resistor</u>. It is often necessary to limit the amount of <u>current</u> in a circuit to avoid the risk of fires. [5]

Answers

1. a.

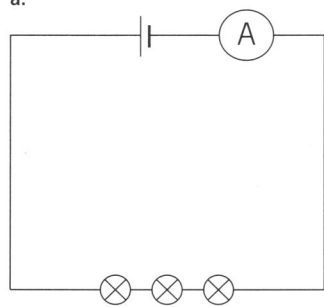

   [Circuit diagram: cell, ammeter A, and three bulbs in series] [3]

   b. The potential difference of the supply/ the type of bulb. [1]
   c. When number of bulbs is doubled from 1 to 2, the current approximately halves from 2.00 to 0.98 A. [1] When the number of bulbs doubles from 2 to 4, the current approximately halves from 0.98 to 0.51 A. [1] This supports the student's conclusion. [1]
   d. As the number of bulbs increases, the resistance in the circuit increases. [1] Doubling the number of bulbs doubles the resistance [1] which halves the current. [1]

Unit 13.3

LL independent variable — the variable that is changed in an experiment [1]

dependent variable — the variable that is measured in an experiment [1]

controlled variable — the variable that is kept the same in an experiment [1]

1. b. potential difference of the cell/resistance of the resistors [1]
   c. 0.20, 0.40, 0.60, 0.80, 1.00 [2]
   d. labelled axes [2] points plotted accurately [1] line of best fit [1]
   e. As the number of resistors increases, current increases. [1] Current is directly proportional to number of resistors. [1]
   f. Increasing the number of resistors increases the number of 'paths' for current flow. [1] Doubling the number of paths halves the resistance [1] and doubles the current. [1]

Unit 13.4

LL series, parallel, resistor, electromotive force [4]

1. a. 11 Ω, 460 Ω, 25 Ω [3]
   b. $I_1 = 0.5$ A, $V_1 = 2$ V [2]
   $I_1 = 2$ A, $I_2 = 1$ A, $V_1 = 1.5$ V [3]
   $I_1 = 0.1$ A, $I_2 = 0.5$ A, $I_3 = 0.2$ A [3]
   c. The 5 Ω resistor carries largest current. [1] For the same potential difference, [1] current is greatest along the path of least resistance. [1]

Unit 13.5

LL Across: 2. cell, 3. battery, 6. light dependent resistor, 8. relay, 10. diode, 11. series [6]

Down: 1. electromotive force, 3. bulb, 4. variable resistor, 5. parallel, 7. thermistor, 9. fuse [1]

1. Input: microphone, LDR, pressure switch, variable resistor, thermistor, reed switch [2]
   Output: loudspeaker, buzzer, heater, LED, lamp, electric bell, relay [2]
2. An analogue signal varies in value continuously. [1] A digital signal has two states: on or off. [1]

Unit 13.6

LL

thermistor	has a resistance that decreases as temperature increases [1]
fuse	melts and breaks the circuit if the current is too high [1]
variable resistor	used to vary the resistance in a circuit [1]
cell	transforms chemical to electrical energy [1]
LED	a diode that emits light [1]
Reed switch	an electrical switch operated by a magnetic field [1]
relay	uses a smaller current to switch on a larger current [1]
LDR	has a resistance that decreases as light intensity increases [1]

1. a. cell [1]
   b. decreases from maximum value at point 1 [1] to zero at point 2 [1]
2. a. cell, diode, bulb [3]
   b. The diode is in reverse bias. [1] The diode should be removed and placed the other way round in the circuit. [1]
   c. A and B [1]
   Current can flow through A because it is in the main part of the circuit [1] and through B because it is in series with a diode in forward bias. [1]

Unit 13.7

LL Logic gates are used to build <u>digital</u> circuits. <u>Logic</u> gates usually have two inputs and one <u>output</u> but the <u>NOT</u> gate has one input and one output. The inputs and output have two possible states: <u>low</u> (0) or <u>high</u> (1), which represent different voltage levels. [6]

1.

Logic gate	Symbol	Truth table
NOT	A ─▷∘─ Y [1]	A | Y 1 | 0 0 | 1 [1]
AND	A ─┐ ⟩─ Y B ─┘ [1]	A | B | Y 0 | 0 | 0 1 | 0 | 0 0 | 1 | 0 1 | 1 | 1 [1]
OR	A ─┐ ⟩─ Y B ─┘ [1]	A | B | Y 0 | 0 | 0 1 | 0 | 1 0 | 1 | 1 1 | 1 | 1 [1]
NAND	A ─┐ ⟩∘─ Y B ─┘ [1]	A | B | Y 0 | 0 | 1 1 | 0 | 1 0 | 1 | 1 1 | 1 | 0 [1]
NOR	A ─┐ ⟩∘─ Y B ─┘ [1]	A | B | Y 0 | 0 | 1 1 | 0 | 0 0 | 1 | 0 1 | 1 | 0 [1]

Unit 13.8

LL Beginnings: endings:

An AND gate gives a high output — if both the inputs are high. [1]
An OR gate gives a low output — if both the inputs are low. [1]
A NOT gate gives a high output — if the input is low. [1]
An OR gate gives a high output — if one or both inputs are high. [1]

1. a. (top) low [1], (bottom) high [1], high [1], low [1]
   b. The output would be high. [1]

Unit 13.9

LL plastic casing, fuse, neutral wire, earth wire [4]

1. a. i. protects the cable from slipping [1] which could cause the earth wire to come loose. [1]
      ii. melts and breaks the circuit if the current becomes too high [1] which protects against fires [1]
      iii. insulates to protect user from electric shocks [1]
   b. If the case of the appliance becomes live [1], a large current flows to earth along the earth wire. [1] The fuse melts and breaks the circuit. [1]

Unit 13.10

LL A mains circuit consists of a live wire and a <u>neutral</u> wire. The <u>potential difference</u> between live and earth alternates between <u>positive</u> and negative. In most countries around the world, the voltage alternates <u>fifty</u> times per second (at a frequency of 50 <u>Hz</u>). The potential difference between neutral and <u>earth</u> is close to <u>zero</u>. A <u>fuse</u> in the live wire protects against fires by melting and breaking the circuit if the current becomes too <u>large</u>. [9]

Answers

1. a. When the current becomes too large [1] in the circuit the magnetic field due to the electromagnet is strong enough [1] to attract the iron rocker and break the circuit. [1]
   b. The device does not need an earth wire [1] because the case is an insulator. [1]

Unit 14.1
LL iron armature, insulator, relay, electromagnet [4]

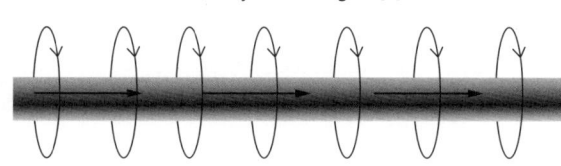

1. [diagram of solenoid with magnetic field lines]

2. When current flows in the circuit with small current and p.d., the coil becomes magnetised. [1] The iron armature is attracted to the coil. [1] The armature swings and closes the high voltage switch contacts. [1]

Unit 14.2
LL When a wire carrying a current is placed in a magnetic field, there is a force on the wire. The size of the force is dependent on the size of the current and the strength of the magnetic field. The direction of the force is dependent on the direction of the current and the direction of the magnetic field. [6]

1. a. Fleming's Left-hand Rule [1] The magnetic field, force and current are at right angles to each other [1] for maximum force on the wire. [1]
   b. i.   move up [1]
      ii. move up [1]
      iii. move down with smaller force/ not as quickly [1]
      iv. move down with more force/ more quickly [1]

Unit 14.3
LL

[word search grid] [8]

1. a. i.   increase [1]
      ii. increase [1]
      iii. decrease [1]
      iv. increase [1]
   b. they provide a constant magnetic field [1]; curved poles help to keep the field radial [1]
   c. Brushes give electrical contact between the external circuit and the rotating coil. [1] They allow the coil to spin and current to flow. [1] Often made from graphite. [1]

Unit 14.4
LL When a magnet is moved into and out of a coil, a potential difference is induced across the coil. The direction of the potential difference changes as the direction of motion of the magnet changes. If there is a complete circuit, an alternating current will flow. [7]

1. a. small negative deflection on the voltmeter, large negative deflection on the voltmeter, large positive deflection on the voltmeter. [3]
   b. all deflections on the voltmeter would be larger [1]
   c. As the magnet moves into the coil, it produces a changing magnetic field in the coil. [1] The changing magnetic field induces a potential difference across the coil [1] because the coil cuts the magnetic field lines. [1]
   d. Positive deflections are produced for positive changes in magnetic field (north pole of magnet moves into the coil) [1] and negative deflections are produced for negative changes in magnetic field (south pole of magnet moves out of the coil). [1]

Unit 14.5
LL

Beginnings:	Endings:
When the coil spins it	cuts the magnetic field lines due to the magnet. [1]
When the coil cuts the field lines, a potential	difference is induced across the coil. [1]
As there is a complete circuit,	an alternating current flows in the coil. [1]
The sides of the coil alternately cut down and	then up through the field lines. [1]

1. a. d.c. flows in one direction in a circuit [1] a.c. constantly changes direction [1]
   b. voltmeter drawn in parallel with the resistor [1]
   c. maintain electrical contact [1] but allow the coil to turn [1]
   d. The voltmeter needle will oscillate [1] from positive to negative p.d. each half turn. [1]

Unit 14.6
LL transformer, primary, secondary, alternating [4]

1. a. $\dfrac{N_p}{N_s} = \dfrac{V_p}{V_s}$ [1]
   b. $12 \times 100 \div 50$ [2] = 24 V [1]
2. a. 1, 5, 3, 2, 6, 4 [6]
   b. The primary coil should have more turns than the secondary coil. [2]
   c. $12 \times 1000 \div 50$ [2] = 240 V [1]

Unit 14.7
LL Across: 3. iron, 6. secondary, 8. electromagnetic induction, 11. more, 12. force [5]
    Down: 1. stronger, 2. transformer, 4. primary, 5. faster, 7. direction, 9. turns, 10. cuts, 13. coil [8]

1. a. For a constant power output from a power station [1], increasing the potential difference decreases the current flowing in the cables. [1] Reducing the current in the cables reduces the heat dissipated into the surroundings [1] and hence increases the efficiency. [1]
   b. Mains electricity is produced in a.c. generators in power stations [1] which means that it can be stepped up using transformers to reduce energy losses on power lines [1] and stepped down to reduce the danger to the consumer. [1]
   c. Advantage: not visible [1] Disadvantage: more costly to install/repair [1]

Unit 15.1
LL

protactinium	a radioisotope [1]
radioisotope	an isotope with unstable nuclei that emit radioactive particles [1]
half-life	the time taken for half of the radioactive nuclei to decay [1]
background count	the contribution of building materials, food etc. to the detected radiation [1]

Answers

1. a. Keep the sources in a lead-lined box until ready to use [1], do not point the source at the students [1], handle the source with long forceps. [1]
    b. 14 [1]
    c. Averaging reduces the effect of random errors [1] and leads to a value closer to the true value. [1]
    c. top row: 319, 312, 7 [3]
       middle row: 327, 150, 177 [3]
       bottom row: 307, 53, 254 [3]
    d. beta, gamma [1]

Unit 15.2

LL radioactive — has an unstable nucleus and emits radioactive particles [1]
    alpha — a helium nucleus [1]
    beta — a fast moving electron [1]
    gamma — an electromagnetic wave of high energy [1]
    ionising — collides with and removes electrons from atoms [1]
    penetrating — can get through many materials [1]
    deflected — changed direction [1]

1. a. alpha, beta, gamma [3]
    b.

type of radiation	alpha [1]	beta [1]	gamma [1]
what is it?	helium nucleus	fast-moving electron	electromagnetic wave [1]
relative charge	+2 [1]	−1	0 [1]
relative mass	4 [1]	1/1840 [1]	0

    c. Because it is unstable [1]
    d. removes an electron/ charges the atom [1]
    e. i. gamma, beta, alpha [1]
       ii. alpha, beta, gamma [1]
    f. top – beta, middle – gamma, bottom – alpha [1]

Unit 15.3

LL alpha particle, gold foil, deflected, radioactive [4]

1. a. i. Alpha particles would be absorbed by air [2]
       ii. to record the number of alpha particles at each angle [1]
       iii. to enable the angles to be measured with greater accuracy [2]
    b. i. An atom's structure consists mainly of empty space. [1]
       ii. Electrostatic repulsion [1] between positive alpha particles and positive gold nuclei. [1]
       iii. Only alpha particles that get very close to the nucleus are strongly repelled. [1] Very few are repelled through 180 degrees and so the nucleus must be very small compared to the size of the atom. [1]

Unit 15.4

LL atom — a tiny particle containing a nucleus and electrons [1]
    electron — a negatively charged particle of very small mass [1]
    proton — a positively charged particle found in the nucleus [1]
    neutron — an uncharged particle found in the nucleus [1]
    nucleus — the central part of the atom [1]
    orbit — the path taken by an electron around the nucleus [1]
    isotope — an atom containing the same number of protons but a different number of neutrons [1]

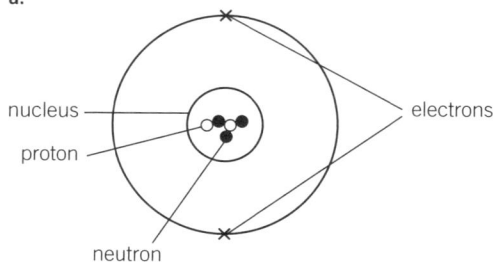

1. a. [diagram of atom labelled nucleus, proton, neutron, electrons]
    b. proton: 1, +1; neutron: 1, 0; electron: $\frac{1}{1840}$ [i.e. negligible], −1 [6]
    c. nucleons [1]
    d. An atom of an element with the same number of protons [1] but different number of neutrons. [1]
    e. Diagram should be identical [1] but with one more/less neutron. [1]

2. a. i. −1 (bottom), 0 (top) [2]   ii. beta particle (electron) [1]
    b. i. 27 (bottom), 60 (top) [2]   ii. cobalt-60 [1]

Unit 15.5

LL A <u>Geiger-Muller tube</u> and a <u>counter</u> are used to detect radiation from a <u>source</u>. Different types of <u>radiation</u> are absorbed by different materials. The <u>background count</u> must be taken first so that this can be <u>subtracted</u> from the count from the source. Radioactive decay is a <u>random</u> process and so it is important to take several readings and <u>average</u> them to obtain a more <u>accurate</u> value. [9]

1. a. i. the time taken for the count rate to halve [1]
       ii. 110−40 [1]   = 70 s [1]
    b. i. 2 × 70 [1]   = 140 s [1]
       ii. 4 × 70 [1]   = 280 s [1]

Unit 15.6

LL proton number, emission, nucleon number, half-life, unstable, nucleus [6]

1. a. Beta radiation is partially absorbed by paper [1]. Gamma radiation is not absorbed by paper at all [1]. Alpha radiation is completely absorbed by the thinnest paper. [1]
    b. The beta radiation is absorbed more than usual [1] and the count rate decreases. [1]
    c. pushes the rollers together [1]

2. a. Gamma [1] since neither alpha or beta can penetrate the soil. [1] Gamma is only able to penetrate the pipe where there is a leak. [1]

Answers to language focus

Exercise 1

1. drag, weight, upthrust (buoyancy)
2. magnetic, electrostatic, gravitational
3. Any three from: chemical, gravitational potential, nuclear, elastic potential (strain)
4. force, moment, speed (velocity), distance (displacement), acceleration, pressure
5. Any three from: hydroelectric, wave, wind, solar, tidal, geothermal, biofuel
6. Any two from: nuclear, geothermal, tidal
7. manometer
8. friction
9. air resistance (drag)
10. density
11. metre rule
12. barometer
13. direct proportion
14. impulse
15. the momentum before a collision is equal to the momentum after collision, provided no external forces act.

Exercise 2

1. states of matter
2. vibrate in fixed positions
3. gases
4. in constant random motion
5. evaporation
6. boiling
7. expansion
8. thermostat
9. gases
10. thermistor
11. internal energy
12. latent heat
13. condensation
14. metals
15. expands, kinetic energy
16. infrared
17. convection current
18. specific heat capacity
19. inverse proportion
20. Kelvin

Answers

Exercise 3
1. wavelength
2. amplitude
3. frequency
4. frequency, speed (velocity)
5. wavefront
6. refraction
7. angle of incidence, angle of reflection
8. real
9. echo
10. Any three from: travel at 300 million m/s in a vacuum; reflect; refract; diffract
11. diffraction
12. dispersion
13. Any three from: virtual, upright, appears to be the same distance behind the mirror as the object is in front, laterally inverted
14. ultraviolet, X-ray, gamma
15. total internal reflection

Exercise 4
1. potential difference, current, charge, energy, power, resistance
2. electric field
3. repels
4. number of turns, iron core, current
5. motor
6. relay
7. light dependent resistor
8. bulb
9. inverse proportion
10. fuse
11. earth wire
12. induced
13. transformer
14. force (motion)
15. commutator
16. slip rings
17. diode

Exercise 5
1. protons, neutrons
2. electrons
3. protons, electrons
4. nucleon (or mass) number
5. alpha
6. fission
7. fusion
8. gamma
9. alpha
10. half-life
11. Geiger counter
12. ionisation
13. film badge
14. gamma
15. beta
16. radioactive tracing
17. kills cancer cells
18. beta
19. concentrated positive charge
20. alpha particles are absorbed by air

Unit 15
1. a. Constant speed [1] of 3.6 m/s. [1]
   d. i. The graph is a straight [1] horizontal line. [1]
      ii. Velocity has magnitude and direction (vector) [1], speed has magnitude only (scalar). [1]
2. a. 1.2 s; discard the anomalous result (1.4 s) and average the rest to calculate a more accurate average. [1]
   b. Start the stop clock at the centre of an oscillation; [1]
      count at least 10 oscillations; [1]
      stop the stop clock; [1]
      divide the total time by the number of oscillations to get a value for the time period of one oscillation. [1]
3. a. $m \times g \times h = 50 \times 10 \times 5$ [1] $= 2500$ J [1]
   b. i. 2500 J [1]
      ii. $v^2 = 2\,KE \div m$ [1] $= (2 \times 2500) \div 50 = 100$ [1]
          $v = \sqrt{100} = 10$ m/s [1]
4. a. combined mass = 300 g = 0.3 kg [1]   weight = 3 N [1]
   b. $p = F \div A = 3(1 \times 10^{-4})$ [1] $= 30\,000$ N/m² [1]
   c. Place the wooden rod in the hole until it reaches the bottom and mark on the rod the level of the top of the sand with a pencil. [1]
      Remove the rod from the hole and the metre rule to measure the distance between the mark and the bottom end of the rod. [1]
   d. 44 mm [1]; doubling the weight will double the pressure [1] and hence double the depth. [1]
5. b. virtual [1] upright [1] magnified [1]
   c. Magnifying glass
6. a. Energy required to raise the temperature of 1 kg of water [1] by 1 °C. [1]
   b. i. 200 [1]   g [1]
      ii. $E = m \times c \times \Delta\theta = 200 \times 4.2 \times 5$ [1] $= 4200$ J [1]
      iii. 4200 J [1]
      iv. $m = E \div (c \times \Delta\theta) = 4200 \div (15 \times 4.2)$ [1] $= 67$ g [1]
      v. $67 \div 1$ [1] $= 67$ cm³ [1]
7. b. diffraction [1]
   c. $f = v \div \lambda = 0.2 \div 0.1$ [1] $= 2$ [1] Hz [1]
8. b. i. 6 V [1]
      ii. $I = P \div V = 10 \div 6$ [1] $= 1.7$ A [1]
      iii. $Q = I \times t = 1.7 \times 3 \times 60$ [1] $= 300$ C [1]
   c. $E = P \times t = 10 \times 3 \times 60$ [1] $= 1800$ J [1]
9. a. i. $N_s = 500 \times 12 \div 230$ [1] $= 26$ turns [1]
      ii. 50 W [1]
      iii. $I = P \div V = 50 \div 12$ [1] $= 4.2$ A [1]
   b. An a.c. potential difference across the primary coil produces an a.c. current in the primary. [1]
      This produces an alternating magnetic field in the primary coil and hence in the core. [1]
      The alternating magnetic field in the core induces a potential difference ad hence a current in the secondary coil. [1]
10. a. i. number of protons [1]
       ii. number of neutrons [1]
    b. A radioisotope has an unstable nucleus which decays by the emission of radioactive particles. [1]
    c. Aluminium: 13 (bottom), 28 (top), Beta particle: −1 (bottom), 0 (top) [4]
11. b.

Inputs		Output
0	0	1
0	1	0
1	0	0
1	1	0

[4]

   c. i. low [1], high [1]
      ii. no effect [1]

Multiple choice answers

Unit 1

1 C   2 B   3 C   4 B   5 C   6 C   7 D   8 B   9 D

Unit 2

1 B   2 B   3 A   4 A   5 A   6 C

Unit 3

1 D   2 C   3 B   4 B

Unit 4

1 B   2 B   3 C   4 D   5 C

Unit 5

1 D   2 D   3 B   4 C   5 C   6 A   7 D

Unit 6

1 B   2 A   3 B   4 A   5 C   6 D   7 D   8 A   9 D   10 A

Unit 7

1 B   2 D   3 C   4 A   5 C   6 B   7 A

Unit 8

1 A   2 B   3 C   4 D   5 C   6 A   7 A   8 B   9 C   10 B   11 A   12 C
13 B   14 C   15 A   16 A   17 D

Unit 9

1 D   2 D   3 B   4 A   5 C

Unit 10

1 C   2 A   3 C   4 D

Unit 11

1 A   2 C   3 B   4 D

Unit 12

1 D   2 A   3 A   4 C

Unit 13

1 B   2 C   3 A   4 D   5 B   6 A   7 B   8 C   9 A   10 B   11 D   12 C

Unit 14

1 A   2 D   3 B   4 D

Unit 15

1 A   2 A   3 B   4 A   5 C   6 B   7 B   8 B   9 B   10 D   11 D   12 C
13 C   14 C   15 C   16 A   17 B

Data sheet

Useful equations

In most cases, the equations below are given in both word and symbol form.

$g = 10 \, \text{N/kg}$ (Earth's gravitational field strength)
$ = 10 \, \text{m/s}^2$ (acceleration of free fall)

Density, mass, and volume
$$\rho = \frac{m}{V}$$

Acceleration
$$a = \frac{v - u}{t}$$

Force, mass, and acceleration
$$F = ma$$

Weight
$$W = mg$$

Stretched spring
$$F = kx$$

Pressure and force
$$p = \frac{F}{A}$$

Pressure in a liquid
$$p = \rho g h$$

Work
$$W = Fd$$

Gravitational potential energy
$$PE = mgh$$

Kinetic energy
$$KE = \tfrac{1}{2} m v^2$$

Energy and temperature change
$$E = mc\Delta T$$

Energy and state change
$$E = mL$$

Waves
$$v = f\lambda$$

Refraction of light
$$n = \frac{\sin i}{\sin r}$$

Total internal reflection
$$\sin c = \frac{1}{n}$$

Charge and current
$$Q = It$$

Resistance, PD (voltage), and current
$$R = \frac{V}{I}$$

Resistors in series....
$$R = R_1 + R_2$$

...and in parallel
$$\frac{1}{R} = \frac{1}{R_1} + \frac{1}{R_2}$$

Electrical power
$$P = VI$$

Electrical energy
$$E = VIt$$

Transformers
$$\frac{V_2}{V_1} = \frac{n_2}{n_1}$$

For 100% efficient transformer:
$$V_1 I_1 = V_2 I_2$$

Data sheet

SI units and prefixes

quantity	unit	symbol
mass	kilogram	kg
length	metre	m
time	second	s
area	square metre	m²
volume	cubic metre	m³
force	newton	N
weight	newton	N
pressure	pascal	Pa
energy	joule	J
work	joule	J
power	watt	W
frequency	hertz	Hz
p.d., e.m.f. (voltage)	volt	V
current	ampere	A
resistance	ohm	Ω
charge	coulomb	C
capacitance	farad	F
temperature	Kelvin	K
	degree Celsius	°C

prefix	meaning	
G (giga)	1 000 000 000	(10^9)
M (mega)	1 000 000	(10^6)
k (kilo)	1000	(10^3)
d (deci)	$\frac{1}{10}$	(10^{-1})
c (centi)	$\frac{1}{100}$	(10^{-2})
m (milli)	$\frac{1}{1000}$	(10^{-3})
μ (micro)	$\frac{1}{1\,000\,000}$	(10^{-6})
n (nano)	$\frac{1}{1\,000\,000\,000}$	(10^{-9})
p (pico)	$\frac{1}{1\,000\,000\,000\,000}$	(10^{-12})

Electrical symbols

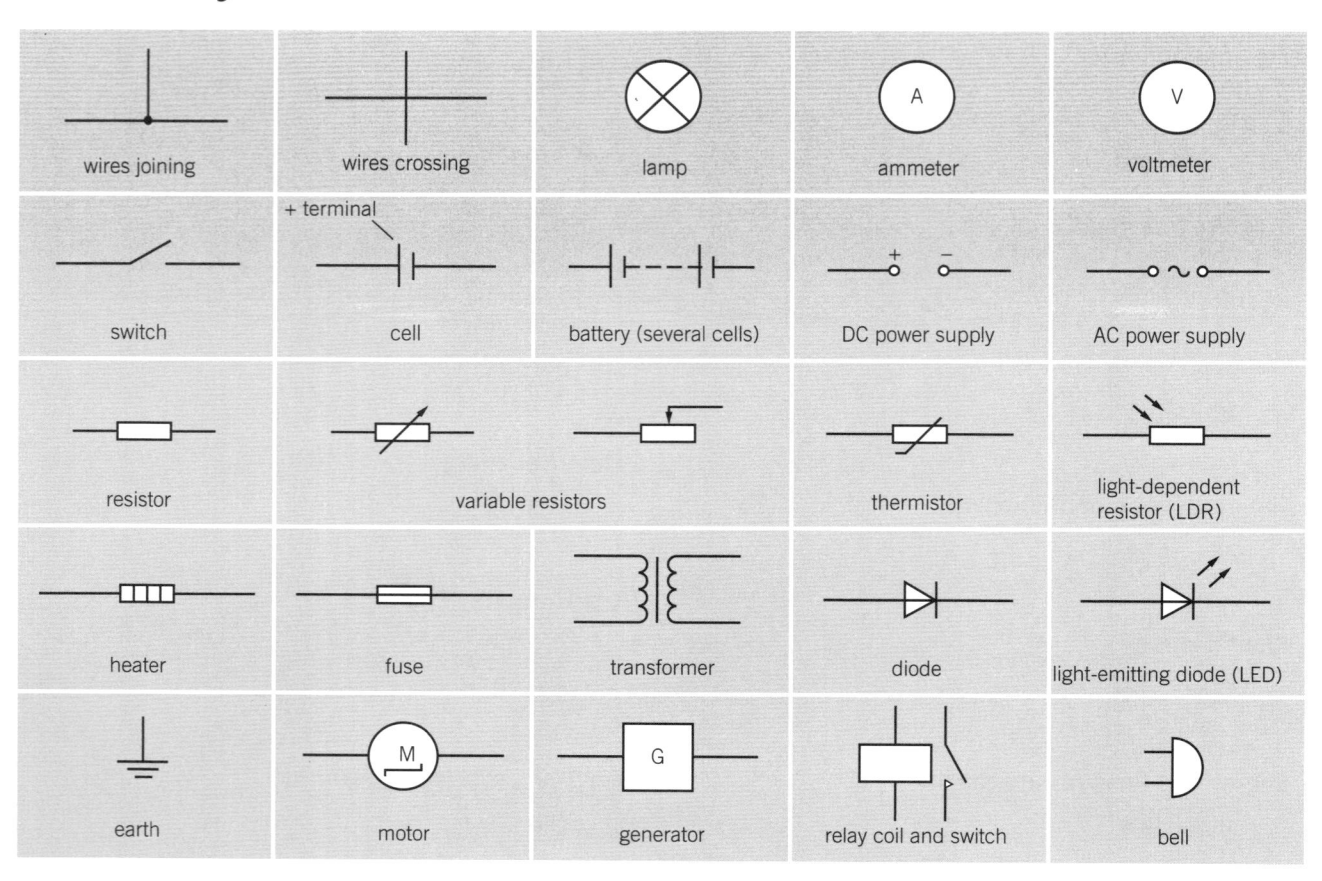

wires joining	wires crossing	lamp	ammeter	voltmeter
switch	cell (+ terminal)	battery (several cells)	DC power supply	AC power supply
resistor	variable resistors	variable resistors	thermistor	light-dependent resistor (LDR)
heater	fuse	transformer	diode	light-emitting diode (LED)
earth	motor	generator	relay coil and switch	bell

172